I0049343

Advanced Applications of Polysaccharides and their Composites

Edited by

Amir Al-Ahmed[1] and Inamuddin[2,3,4]

[1]Centre of Research Excellence in Renewable Energy, King Fahd University of Petroleum & Minerals, Dhahran, Saudi Arabia

[2]Chemistry Department, Faculty of Science, King Abdulaziz University, Jeddah 21589, Saudi Arabia

[3]Centre of Excellence for Advanced Materials Research, King Abdulaziz University, Jeddah 21589, Saudi Arabia

[4]Department of Applied Chemistry, Faculty of Engineering and Technology, Aligarh Muslim University, Aligarh-202 002, India

Copyright © 2020 by the authors

Published by **Materials Research Forum LLC**
Millersville, PA 17551, USA

All rights reserved. No part of the contents of this book may be reproduced or transmitted in any form or by any means without the written permission of the publisher.

Published as part of the book series
Materials Research Foundations
Volume 73 (2020)
ISSN 2471-8890 (Print)
ISSN 2471-8904 (Online)

Print ISBN 978-1-64490-076-5
eBook ISBN 978-1-64490-077-2

This book contains information obtained from authentic and highly regarded sources. Reasonable efforts have been made to publish reliable data and information, but the author and publisher cannot assume responsibility for the validity of all materials or the consequences of their use. The authors and publishers have attempted to trace the copyright holders of all material reproduced in this publication and apologize to copyright holders if permission to publish in this form has not been obtained. If any copyright material has not been acknowledged please write and let us know so we may rectify this in any future reprints.

Distributed worldwide by

Materials Research Forum LLC
105 Springdale Lane
Millersville, PA 17551
USA
https://www.mrforum.com

Manufactured in the United States of America
10 9 8 7 6 5 4 3 2 1

Table of Contents

Preface

Keyword Index

About the Editors

Preface

Polysaccharides are natural polymers and have huge application potentials in many of our daily utilities especially in the medical and food sectors. These materials have vast economic potential and are at the same time environment friendly and nontoxic. In some cases pure polysaccharides suffer from poor mechanical properties. Thus polysaccharide-based composites are more attractive and extensively studied. Many fibers and friendly nano materials are reinforced in polysaccharides to prepare composites and have desired application properties. Thus, polysaccharide-based composites are one of the best alternatives to non-biodegradable petroleum-based polymers and getting increasing application clearances. Due to its huge prospect many researchers are engaged to come up with advanced materials and novel application ideas.

With the progress of technologies, concepts and penetration of nanotechnology, we are expecting better performing polysaccharide-based for drug delivery or tissue engineering, or food/ pharmaceutical packaging and many other biomedical applications. Due to huge research activity and constant progress on polysaccharide-based composites, this book has been compiled, where we have received chapters from some famous researchers in this sector from around the world who have agreed to share their research expertise as well as visions for the future development of polysaccharide-based composites.

We want to express our gratitude to all the contribution authors, publishers, and other research groups for granting us the copyright permissions to use their illustrations. Although every effort has been made to obtain the copyright permissions from the respective owners to include citation with the reproduced materials, we would like to offer our sincere apologies to any copyright holder if unknowingly their right is being infringed. One of the editors, Dr. Amir Al-Ahmed is tankful to the director of Centre of Research Excellence in Renewable Energy at King Fahd University of Petroleum & Minerals, Saudi Arabia, for his continuous support. Finally, we would like to acknowledge the sincere support of Mr. Thomas Wohlbier of Materials Research Forum LLC in evolving this book into its final shape.

Amir Al-Ahmed
Centre of Research Excellence in Renewable Energy, King Fahd University of Petroleum & Minerals, Dhahran, Saudi Arabia

Inamuddin
Chemistry Department, Faculty of Science, King Abdulaziz University, Jeddah 21589, Saudi Arabia
Centre of Excellence for Advanced Materials Research, King Abdulaziz University, Jeddah 21589, Saudi Arabia
Department of Applied Chemistry, Faculty of Engineering and Technology, Aligarh Muslim University, Aligarh-202 002, India

Advanced Applications of Polysaccharides and their Composites   Materials Research Forum LLC
Materials Research Foundations **73** (2020) 1-26   https://doi.org/10.21741/9781644900772-1

Chapter 1

Polysaccharide Composites as a Wound-Healing Sponge

Ayesha Khalid[1], Naveera Naeem[1], Taous Khan[2] and Fazli Wahid[1,3,*]

[1]Department of Biotechnology, COMSATS University Islamabad, Abbottabad Campus 22060, Pakistan

[2]Department of Pharmacy, COMSATS University Islamabad, Abbottabad Campus 22060, Pakistan

[3]Department of Biomedical Sciences, Pak-Austria Fachhochschule: Institute of Applied Sciences and Technology, Mang, Khanpur Road, Haripur, Pakistan

*fazliwahid@cuiatd.edu.pk

Abstract

The first wound treatment was designed five millennia ago. Since, then tremendous scientific and technological advancements have been made to design multifunctional wound healing materials. This chapter gives a comprehensive discussion on the molecular and cellular events of wound repair and the biological aspects of polysaccharide-based sponges as a dressing system. Overall, this chapter focuses on the polysaccharides including alginate, cellulose, chitosan, hyaluronic acid, dextran and others. It briefly recaps the recent progress made in developing their composite sponges. Furthermore, the healing performance of these sponges is also discussed in detail.

Keywords

Wound Healing, Polysaccharide, Sponges, Dressing, Composites

Contents

1.   Introduction

Human skin is the largest multifunctional organ of the body that acts as a critical barrier between human body and the external environment. The primary function of the skin is protection from physical, chemical and biological assaulters and UV radiations [1]. Other important functions of the skin include thermoregulation, sensation and synthesis of vitamin D. As the skin has so much important functions, therefore, any damage or break in the continuity of the skin must be rapidly and efficiently restored [2]. Naturally, skin has remarkable regenerative properties. Following any injury, epithelial wound healing is an inherently controlled multicellular process comprising of a series of highly coordinated intercellular and intracellular pathways aiming to restore the epidermal structure and function [3]. Acute cutaneous injuries usually heal within a week or two but chronic and infected wounds do not heal in a predictable amount of time therefore necessitate timely treatment and rehabilitation [2].

History of wound management illustrated that how the wound healing products evolved over the years from topical ointments such as animal oils or fat, honey, mud and wine to traditional dressings mainly cotton and wool gauzes. Although, most of these products provided some benefit to acute wounds but fail to cure chronic and hard to heal wounds. Therefore, gauze and cotton dressings were rapidly replaced in the second half of the 19th century. Winter (1962) gave the principle of moist wound healing [2]. His concept sets the ground for the development of new generation of dressings called occlusive dressings including hyrocolloids, films, hydrogels, foam and sponges. These dressings are meant for the functional recovery of wounds by providing a moist healing environment and act as a barrier against microbial function. Semi occlusive and occlusive dressings are usually formulated by different natural polysaccharides due to their high hydrophilicity, biocompatibility and low or no toxicity. In mechanism, polysaccharides target every phase of wound healing thus maximizing the outcome in terms of tissue regeneration and wound healing. Numerous advances have been made in terms of fabrication of these dressings by using a polysaccharide alone or in combination with other synthetic polymers, antimicrobial and wound healing agents to impart additional physicochemical characteristics depending on the nature and type of the wound [2].

This chapter describes the mechanism of wound healing and the use of different polysaccharide for wound care applications with a prime focus on polysaccharide based sponges. The chapter presents the sponges of alginate, cellulose, chitosan, hyaluronic acid, dextran and other with a brief recap of the recent progress made in developing their composite sponges.

2. Wound healing from hemostasis to remodeling

The wound recovery process is initiated by the formation of fibrin clot that shields the skin opening, followed by the regeneration of lost cells resulting in a healed tissue [4]. This complex process involves coordinated action of various cell types like leukocytes, keratinocytes, endothelial cells, platelets, macrophages and fibroblasts. The cells migrate to the wound interface and initiate the recovery process, which is directed and controlled by the equally complex signaling molecules such as growth factors, chemokines and cytokines [5]. These mediators are bio-active polypeptides that regulate the cellular processes by altering the growth, differentiation and metabolism of the target cell. The growth mediators trigger the cell lineages lying at the wound margin by binding to their specific cell surface receptors or extra cellular matrix (ECM) proteins. This binding activates the attachment of transcription factors to the promotor region of the gene that results in the transcription of gene that is translated to a specific protein. The wound healing proteins control the cell cycle, migration and differentiation patterns of the cells in order to fill the wound gap [6]. These molecules actively control all the phases of wound healing. The function of growth regulators are listed in Table 1. Any imbalance in their expression leads to chronic wounds like diabetic foot ulcer, pressure ulcer, and chronic venous leg ulcer [7].

Table 1: Growth mediators expressed by different cell type that are involved in thedifferent phases of healing.

S. No.	Stage of wound healing	Cell type	Wound healing mediators	References
1.	Hemostasis	Platelets	PDGF, TGF-β, TGF-α, VEGF, IGF-1 and bFGF	[3,5,6]
2.	Inflammatory	Monocytes	IL-1, IL-6, IL-8, PDGF, TGF-α, TGF-β	[7,10]
		Neutrophils & Macrophages	TNF-α, IFN-γ, IGF-1 and FGF	
3.	Proliferative	Fibroblasts	IL-1, PDGF, TGF-α, TGF-β, IGF-1 and EGF	[3,5,7]
		Endothelial cells	PDGF, VEGF, IGF-1, bFGF and EGF	
		Keratinocytes	EGF, IGF, TGF-α, KGF and bFGF	
4.	Remodeling	Fibroblasts	PDGF, TGF-β, EGF and bFGF	[6,7]

Wound healing comprises of chemically controlled four phases that overlap in space and time; hemostasis, inflammation, proliferation and remodeling. Within a few moments after an injury, the process of hemostasis gets activated to restrict the outflow of blood. The hemostasis is characterized by vasoconstriction proceeded with platelet activation. Activated platelets aggregate and get attached to the free collagen surfaces in the ECM in order to form a platelet plug. This clot is composed of interlinked fibrin, vitronectin, fibronectin, thrombospondin, erythrocytes and platelets, which serves as a provisional matrix. The fibrin matrix befits like a scaffold in order to support infiltrating cells and provide barrier against the invading microbes [8]. Additionally, this matrix acts as a reservoir of chemical mediators necessary for the later stages of wound healing [9]. Followed by hemostasis, inflammatory cells recruit to the wound bed, initiating the next phase of healing called inflammation. This phase functions in two ways by removing dead cells and invaded microbes followed by deposition of collagen and neovascularization. Here, mast cells, neutrophils and macrophages are actively involved [10]. After 5 to 10 min of vasoconstriction, mast cells stimulate vasodilation by releasing vasoactive amines and histamine rich granules, helping in migration of inflammatory cells at the wound interphase. Neutrophils and monocyte act as a scavenger cell in the inflammatory phase of healing. The migration of both cells from the circulating blood into extravascular space is facilitated by the slight molecular changes in the endothelial cell surface [11]. Initially, slight interaction of leukocyte with the members of selectin family of adhesion molecules allow them to slowdown and drag from the rapid circulation, then considerable adhesion is mediated by integrins $\beta2$ which lead to extravasation of neutrophils from the blood stream to the wound site. Within minutes, active neutrophils arrive at the injury site to clear the invaded contaminants. Along with that, neutrophils produce specific signals in the form of pro-inflammatory cytokines to activate cells involved in the next phase of healing. As the wound bed gets cleared, neutrophil milieu is phagocytosed by the tissue macrophages. Macrophages continue to gather at the injury site by differentiation of monocytes released by blood stream. Macrophages remove any pathogenic remnants and matrix debris. Moreover, once activated, macrophages create signals for transition into proliferation phase such as, collagen production, re-epithelialization and angiogenesis [8]. The proliferative phase lasts 2 to 10 days post injury, characterized by migration and proliferation of various cell types. The epidermal growth factor regulates the infiltration of keratinocytes at the wound bed. New blood vessels formed by a process called angiogenesis to fulfil the nutrient demand of the new skin cells [12]. Subsequently, fibroblast and macrophages migrate and replace the fibrin matrix with granulation tissue [3]. In the later part of this phase, macrophages stimulate fibroblasts lying at the wound margin or bone marrow to differentiate into contractile cells called myofibroblast. Fibroblast and its differentiated phenotype produce the collagen (Type-III) which is the main component of scar tissue. The last phase of healing includes rearrangement of new cells to mimic the normal anatomy of native skin. Remodeling starts after 2 to 3 weeks of injury and lasts for a month or more [13]. It is the longest phase of healing characterized by dynamic changes in tissue architecture by the involvement of proteolytic enzymes, mainly matrix metalloproteinases (MMPs) and tissue inhibitors of metalloproteinases (TIMPs). These enzymes function in maintaining balance between synthesis and breakdown of ECM [14]. Most of the proliferative reactions ceases at this stage followed by apoptosis of endothelial cells, myofibroblasts and macrophages. As the remodeling continuous epithelial-mesenchymal interactions regulate the skin integrity and homeostasis. The changes in the wound result into a mature wound with a gradual accumulation of type I collagen along with larger blood vessels. Grossly, the healed skin tissue regain 80% of elasticity in comparison with the original skin tissue [15]. Abnormal expression of any of the above mentioned events or growth mediator expression lead to impaired healing of the wound that need treatment.

3. History of wound treatment

Development of treatments for acute and chronic wounds with rapid and effective outcomes remain a challenge since early times. The treatment of skin wounds aims to restore the morphology and functionality of skin. For the injury to heal properly, numerous curative products are designed, ranging from simple coverage, antiseptic solutions, gauzes, gels and ointments to more complex dressing types termed as bioactive or smart dressings [8]. Historically, various treatment strategies were opted to cure wounds of different types. In 1600 BC, linen strips dipped in grease or oil covered with plasters was the common strategy to cover the wound [16]. About 2500 BC, people in Mesopotamia cleaned injury with milk or water followed by application of honey or resin dressing. Furthermore, use of clay tablets was one of the oldest therapies, applied in the form of plasters and bandage. Plasters still finds promising use in the present-day treatment techniques as it provide protection and absorb wound exudate [17]. In 460-370 BC, Hippocrates of ancient Greece washed the wounds with wine or vinegar followed by honey, oil or wines based bandage. In Mesopotamian culture a well-known saying stated "*Pound together fur-turpentine, pine-turpentine, tamarisk, daisy, flour of inninnu strain; mix in milk and beer in a small copper pan; spread on skin; bind on him, and he shall recover*" [18]. Egyptians' art of wrapping mummies introduced the concept of bandaging of wounds. In addition, Egyptians controlled infections by painting green paint on the wound surface, as per their belief that green indicated life but actually the paint contained copper that possess healing activity. Moreover, Greeks used boiled wine or wine bandage to avoid bacterial infection [19]. Later in the 19th century, the introduction of antibiotics bought a major breakthrough in the antiseptic technique that controlled mortalities due to infection [20]. At that time, traditional dressings including gauze bandages made of woven or non-woven fibers of rayon, cotton, polyester and flax were introduced. Gauze dressings soaked in different antiseptic chemicals, such as iodine and metal solutions, were used to control the infections [21]. Gauze dressings were suitable for protecting wounds from contamination and to stop bleeding but they do not play an active role in healing of the damaged tissue. Similarly, these dressings may stick to the wound surface thus desiccate the wound, and cause damage to the regenerated epithelium on removal. Uncovered and dry wounds heal more quickly was the general concept of healing till the mid 19th century [22]. However, later on in the 20th century, the concept of modern wound dressings were introduced when George Winter explains the idea of moist healing. According to his findings published in Nature in 1962, the rate of epithelization in partial thickness wound increases 50% when the wound is covered with polyethylene film as compared to open and dry wounds. Later on, the polyester and polypropylene films were also tested that confirm his idea that moist environment favors faster healing.

These studies eventually change the face of wound management and led towards the development of modern dressing system that provide optimal environment for efficient and proper healing [23].

4.   The need for modern dressings

Loss of integument, excessive fluid loss, polymicrobial infection and inefficient conventional dressing materials are some of the major causes of poor and delayed healing. Dermal wounds are the favorable niche for the colonization of microbes that contaminates the wound and halt the healing process. Therefore, scientists tried to formulate and fabricate multifunctional dressing materials to cater the issues of wound repair process [22]. The ideal dressing should be biocompatible and nontoxic. It must have high water holding capacity for the provision of moist environment to the wound and to absorb excessive exudates. It should act as a barrier against microbial penetration and also have the ability to kill the microbes. Furthermore, the dressing should be non-adherent, easily removable, reduce pain and distress associated with frequent dressing changes [18]. Researchers have designed various forms of dressing materials such as hydrogels, hydrocolloids, films and sponges, with the desire to provide an optimum healing environment.

Polymeric films provide a moist healing environment. These films are usually semipermeable allowing the oxygen to pass through but inhibit the bacterial penetration. The film based dressings provide the occlusive healing environment but do not have the ability to absorb the exudate, which is a major drawback [24]. Hydrocolloids are another class of dressing generally made up of hydrophilic polysaccharides. Hydrocolloids are occlusive in nature, aids the autolytic debridement, and highly absorbent in nature, therefore, widely recommended for the treatment of chronic wounds such as diabetic ulcers. Hydrogels are the cross-linked hydrophilic polymers such as polyacrylamide and polyethylene oxide that do not dissolve in water but have high water holding and swelling properties. However, hydrogels have weak mechanical properties, and therefore, an additional dressing is required [22].

5.   Sponges

Sponges are the soft and flexible scaffolds with three-dimensional, interconnected porous structure. The highly porous unique architecture of sponges closely mimics the extracellular matrix and promote the cellular interaction, adhesion, growth and proliferation. Moreover, their ability to retain moisture helps in sustaining a moist environment that is ideal for the healing of wounds [25]. The micro porosity in sponges allow absorption of large amount of fluid, approximately, more than 20 times the dry weight of the sponge. This property helps in wound exudate absorption (Fig. 1).

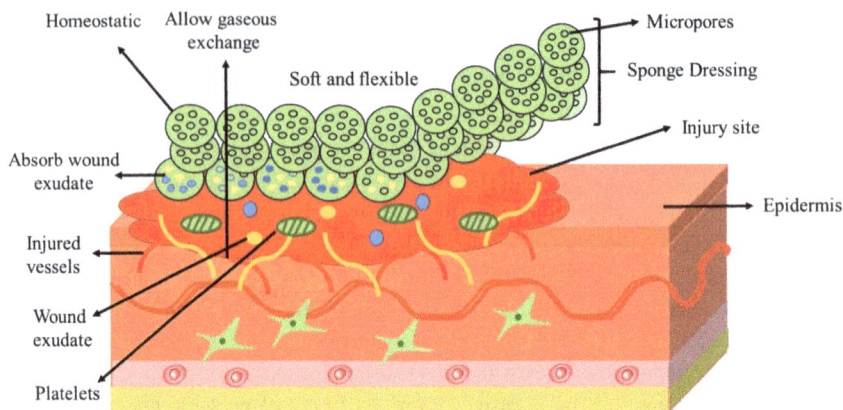

Figure 1. Basic features of sponge and its healing properties. High porosity allow gaseous exchange while microporous structure limits microbial penetration. The swelling property helps in absorbing wound exudates and homeostatic ability prevent fluid and blood loss from open wound. Water retention features provide moist environment that favors autolytic debridement and promote cellular migration and proliferation.

Nowadays, biopolymers find promising applications in biomedical fields particularly tissue engineering. Polymeric blends improve the characteristic properties of the polymers, finding new treatment options for wound management. Therefore, various forms of polysachharide based dressings have been introduced to effectively combat infections and enhance healing activities [26].

6.   Wound healing potential of polysaccharides

Polysaccharides have been largely explored as a dressing material for a wide range of wounds in the last decade. Various synthetic and natural polysaccharides and their combinations are receiving attention from scientists worldwide for the fabrication of advancing the wound care dressings. The unique properties of natural polysaccharides such as excellent biocompatibility, antimicrobial property, non-cytotoxicity, similarity to extra cellular matrix and homeostatic ability make them a superior choice for designing of efficient wound care material. The polyaccharides are abundantly found in different plants and animals [27].

Polysaccharides containing a β-glucan linker contribute to the healing process by stimulating the immune system involving activation of macrophages. Macrophages are one of the important inflammatory cells involved in debridement of the wound, resolution of inflammatory phase, removal of apoptotic cells, promote initiation of proliferative phase and tissue restoration followed by any injury. Several studies reported that mechanistically, natural polysaccharide can enhance cytotoxic activity of macrophage against microbes and tumor cells along with that, escalate reactive oxygen species (ROS) and nitric oxide (NO) production in order to activate its phagocytic activity. These aptitudes are useful in achieving better healing in lesser time. [28]

In addition, various studies have explored that the role of natural polysaccharides in the modulation of wound healing related pathways. For example, natural polysaccharide, by its oligosaccharides (xyloglucan, β-glucan, chitin, D-mannuronic acid, pectin and L-guluronic acid) activate the cells to produce cytokines involved in healing. Principally, β-glucan is known to exert its effect by binding to several receptors like Toll-like receptor (TLR), scavenger receptor, dectin-1 receptor and lactocylceramide [29, 30]. In one mechanism, β-glucan binds with the dectin-1 receptor and stimulate the production of cytokines or activates other inflammatory and non-inflammatory reactions [31]. In an experiment, polysaccharide-rich fraction of *Agaricus brasiliensis* modulated the pro- and anti-inflammatory reactions by interacting with the TLR 2 and 4, thereby increasing production of tumor necrosis factor alpha (TNF-α) and interleukin-1β (IL-1β) by the monocytes [32]. TNF-α is an important agent in inflammatory phase as it is involved in chemotaxis of neutrophil. While, IL-1β is a critical inflammatory mediator, known to control neutrophil mobilization, adhesion to endothelial wall and infiltration of leuckocytes [33, 34]. In another study, polysaccharide form *Astragalus membranaceus* was used for wound healing activity. The results showed that the respective polysaccharide enhance fibroblast propagation, cytokine production, revascularization and reepithelization [35]. In addition, a *Sanguisorba officinalis L.* polysaccharide stimulated the healing process by enhancement of collagen synthesis and angiogenesis by the activation of IL-1β and vascular endothelial growth factor (VEGF) [36]. Of all the polysaccharides, alginate, hyaluronic acid, chitosan and cellulose are of prime importance. These polysaccharides are widely investigated for fabricating variety of wound care products alone or in combination with other polysaccharides and wound healing agents as composite products.

6.1 Alginate composite sponges

Alginate, also called alginic acid or algin, is an important anionic polysaccharide, widely distributed in the cell walls of brown seaweeds including *Ascophyllum, Ecklonia, Turbinaria, Sargassum, Macrocystic, Durvillaea* and *Laminaria* species. Chemical structure of alginate include linear unbranched chain of two polymeric residues β-(1-4) linked d-mannuronic acid and β-(1-4) linked l-glucoronic acid as shown in Fig. 2. Alginate is an eco-friendly polymer with controllable porosity. Alginate and its salts found applications in wound healing due to its hemostatic properties, furthermore, it can promote cell adhesion and proliferation of new skin cells. It has the capability to absorb body fluids up to 20 times of its weight, and thus is useful in designing dressings for highly exudating wounds [37]. Similarly, alginate is rich in glycosaminoglycan (GAG) content, which is well-known to achieve faster healing by rapid granulation, angiogenesis and regeneration of epithelia. This ultimately results in healed skin with minimum scaring [38]. GAG is an extracellular matrix molecule having a vital role in wound healing either acute or chronic [39]. In addition, once alginate based dressing get attached to the wound site, an ion-exchange reaction starts between calcium of alginate and sodium from the exudate, producing soluble gel that help in maintenance of moisture. Thus, moist environment facilitate the migration and growth of cells involved in healing [40]. Similarly, after injury body fluid and blood loss takes place from the injured blood vessels. In order to limit such loss, a hemostatic sponge was designed using sodium-alginate and carboxymethyl chitosan as the main material. The result of an *in-vitro* assay displayed procoagulant activities of the composite [41].

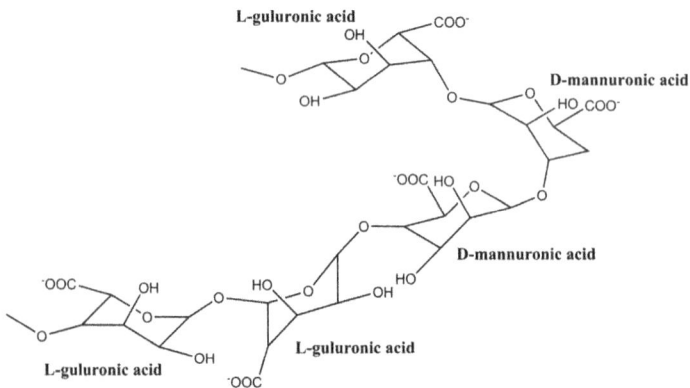

Figure 2. Chemical structure of alginate that constitues L-guluronic acid and D-mannuronuc acid units in linear unbranched chain form.

In a pre-clinical study, calcium alginate dressing enhanced the expression of transforming growth factor family (TGF-β1, TGF-β2 and TGF-β3), suggesting a role in the proliferative stage of healing. Likewise, alginate and silk fibroin (sponge) was individually applied on the full thickness rat wound model. The blend of silk fibroin/alginate was applied to evaluate the synergistic effect. The fibroin/alginate treatment was superior amongst other, promoting re-epithelization by enhanced proliferation of epithelial cells [42]. Therefore, the healing potential of alginate finds potential application in designing an ideal dressing system.

6.2   Cellulose composite sponges

Cellulose is the most abundant natural polysaccharide present on earth. Its chemical structure comprised of a linear chain of β 1-4 D-glucose units as shown in Fig. 3. These glucose units aggregate with each other to form thread like fibrils. These fibrils combine themselves in the form of bundles called microfibrils that form a three-dimensional matrix. It is an important structural component of the plant cell wall and also produced by some strains of bacteria. Cellulose obtained from both sources is chemically similar with slight differences [43]. Bacterial cellulose (BC) is the purest form of cellulose and is far superior to plant cellulose as it does not possess any impurities like pectin, lignin and hemicellulose. Due to the unique three-dimensional macro and micromorphology of BC, it possesses distinctive physical and mechanical properties that make it an ideal material for biomedical applications such as in tissue engineering, drug delivery, regenerative medicines, dental and bone implants, artificial skin and cutaneous wound dressings [44]. Among these, the use of BC for the fabrication of wound dressings showed quite promising results. The first successful use of BC for burn wounds and ulcers was documented by a Brazilian company in a series of patents in 1980s. Extensive studies on biocompatibility and biomedical applications of cellulose have been conducted in recent years [44]. The BC has high porosity, permeability and high-water holding capacity. The microporous structure of cellulose helps in controlling the rate of water absorption and water loss thus provides a moist healing environment making it an excellent dressing material. These properties further help in absorbing wound exudates and also allow the exchange of oxygen. BC is a highly elastic material and can be produced and molded into desired shapes. In addition, high crystallinity, good mechanical strength, biocompatibility and nontoxicity are some of the distinctive features of this biopolymer [45].

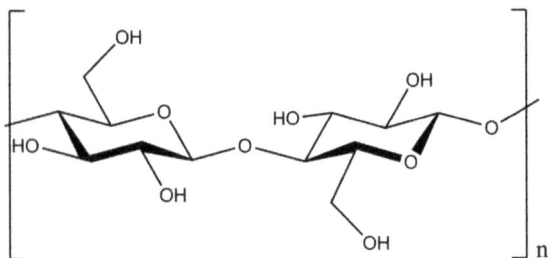

Figure 3. Chemical structure of cellulose having β 1-4 D-glucose units linked by glycosidic bond in a linear chain form.

However, some of the characteristics such as lack of intrinsic antimicrobial activity and non-bioabsorbability limit the medical application of BC. Therefore, polymeric composites of BC are widely fabricated aiming to enhance the biological properties of this multifunctional polymer. Different composites of BC with nanoparticles, biological and synthetic polymers and antimicrobial agents showed huge potential in wound healing and tissue regeneration. A novel biocompatible composite sponge was fabricated by functionalizing regenerated BC (RBC) with amoxicillin (an antimicrobial agent) that possessed enormous potential to control infection and enhance healing [46]. Due to the presence of broad-spectrum antibiotic, amoxicillin, the composite sponge showed superior antimicrobial activity against *Escherichia coli*, *Staphylococcus aureus* and *Candida albicans* as compared to non-functionalized sponge. Furthermore, the composite sponge showed biocompatibility towards HEK293 cell line and also displayed enhanced healing of excisional wound *in-vivo*. [46]. In another study, silver nanoparticles were immobilized in the freeze-dried cellulose matrix to create an antimicrobial composite sponge. *In-vivo* testing revealed that the composite sponge accelerated the healing of infected wounds. [47]

The structural morphology of BC favors the cellular proliferation, growth and adhesion, thus act as a perfect scaffold for tissue regeneration. In one such study, the composite sponge of BC and silk fibroin was fabricated with improved regenerative properties. Findings of this study depicted that the addition of silk fibroin enhances the cellular adhesion and biocompatibility of the BC making it a better candidate for synthesizing tissue regenerating scaffold [48].

The composite sponge of BC and gelatin was fabricated as a multipurpose scaffold for biomedical applications such as tissue engineering and wound healing. The composite sponge showed good compatibility with Vero cell lines. Moreover, the sponge possessed good structural stability in water and was highly porous, thus, mimics the extracellular matrix. Furthermore, the composite sponge displayed faster healing of full thickness wound as compared to gauze dressing [49].

Due to high mechanical strength and flexibility of the nanofibers of BC, different materials were blended to form composites of versatile properties. These composite sponges possess immense healing potential and can be considered as better options for developing wound care treatment.

6.3   Hyaluronic acid composite sponges

Hyaluronic acid (HA), also called hyaluronan, is a natural polymer present in the epithelial, connective and neural tissues of the human body [50]. After its discovery in 1934, this polymer has been widely studied, finding promising applications in the biomedical field. It is a member of a large family called glycosaminoglycan [51]. Chemically, HA polymer is composed of two basic units, D-glucoronic acid and *N*-acetyl-glucosamine joined by β-glycosidic linkages as shown in Fig. 4. Studies revealed that healing properties of HA depends on its molecular size. Low molecular weight HA displays pro-inflammatory properties whereas, high molecular weight HA is a potent inflammatory inhibitor and immune suppressor. In a study, a blend of both types of HA along with collagen containing EGF was prepared in the form of a spongy sheet. The dressing applied on excisional wound of Sprague-Dawley (SD) rat displayed scar free healing. The results define angiogenic, anti-inflammatory and cell migratory role of HA [52]. Similarly, in another study, HA, arginine and EGF based spongy dressing confirm the healing potential of low and high molecular weight of HA. The *in-vivo* experiment indicated moderate inflammation by the application of HA sponge [53]. Several studies confirmed that TLR signaling pathway is involved in HA induced pro-inflammatory reactions. Using *in-vitro* studies, the researchers confirmed that low molecular weight HA bind with TLR, followed by initiation of signaling cascade, production of chemokines and pro-inflammatory cytokines in different cell types [50]. The important biological property of HA is its hygroscopic nature, allowing it to swell 1000 times more in volume. In skin tissues, this feature allows to regulate tissue hydration during the inflammatory process after the damage [54].

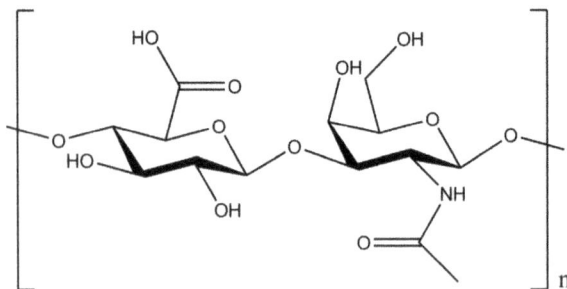

Figure 4. Molecular structure of Hylauronic acid comprising two basic units, D-glucoronic acid and N-acetyl-glucosamine joined by β-glycosidic linkages in a linear form.

Researchers have exploited the wound healing potential of HA to design various formulations aiming to restore the normal skin anatomy and functionality [55]. Recently, HA and collagen sponge materials were developed in various ratios of pig skin and bird feet collagen. The composite may work as artificial skin as it possesses high stability, mechanical strength and healing potential [55]. Another study was conducted with an aim to reduce scaring after breast surgery by combining the individual properties of HA and zinc. The results of the preliminary study showed reduced scaring in patients treated with a zinc-hyaluronan sponge in comparison with the placebo. Histological studies confirmed that reduced scaring is due to changes in the density and orientation of collagen fiber [56]. Scientific studies demonstrate HA as a healing molecule, therefore, a combination of HA, collagen and fibronectin was opted to prepare a wound healing sponge [57]. The *in-vivo* testing showed infiltration of the sponge with the fibroblast leading to increased synthesis of collagen. The positive results of the prepared sponge suggest pre-clinical evaluation on an animal model [57]. As discussed, HA can be used in the future as an advanced dressing material compared to traditional treatments.

6.4   Chitosan composite sponges

Chitosan is a natural copolymer of D-glucosamine and *N*-acetyl-D-glucosamine linked by β (1-4) glycosidic linkages as dipicted in Fig. 5. It is a deactylated product of chitin, obtained from the cell wall of crustaceans and some fungi. The functional groups of chitosan allow surface modifications providing additional bulk properties and enhanced bio-functionality. Chitosan has a wide range of applications in wound dressings, tissue engineering, drug delivery, gene delivery and hemodialysis [58]. A number of studies were conducted to design chitosan based dressings aiming at superior healing outcome. A chitosan sponge is able to provide suitable conditions for effective healing process [59]. A study reported a biodegradable sponge composed of chitosan and sodium alginate fabricated with curcumin. This sponge was ethereal and pliable and applied on a full-thickness wound of a rat. Histological results of gauze treated wounds displayed immature granulation tissue with numerous inflammatory cells and congested vessels, while, well aligned collagen and healthy granulation tissues were observed in wounds treated with prepared sponge. This portrays that the wound was provided with a better healing environment by the sponges [60]. In another study, an antimicrobial chitosan based sponge dressing was designed, to obtain superior healing. A green and facile method was used to obtain ampicillin fabricated chitosan sponges. The sponge was non-leaching, thus, limiting the usage of antibiotic. Besides, the antimicrobial and tissue regenerative capability, the sponge was evaluated *in-vitro* against common microbes and *in-vivo* wound model. The result showed good antimicrobial property with accelerated wound healing competency. Consequently, the chemically enhanced chitosan sponges unveil great potential as promising antimicrobial wound dressings [61]. Another chitosan polymeric sponge formulation was designed containing gelatin followed by crosslinking with tannin. In a 15 days experiment conducted on rabbits, full thickness wound covered with the sponge showed superior healing. Additional loading of the sponge with platelet-rich plasma further enhanced the healing outcomes, which can be used in treating injuries [62]. Moreover, a flexible three- dimensional porous chitosan sponge containing halloysite nanotutes (HNTs) were prepared and applied on full-thickness excision wound in experimental rats. The results showed increased wound closure with 3.4-21 fold change relative to pure chitosan. The addition of HNTs promote re-epithelization and collagen deposition. The chitosan-HNTs composite sponge may have potential application in the treatment of burn wounds, diabetic foot ulcers and chronic wounds [63]. Chitosan based antimicrobial sponge find importance in controlling infection at the wound site. In a study, chitosan, HA and nano silver based wound dressing was designed in the form of sponges to evaluate its healing potential for diabetic foot ulcers. The designed sponge showed good antimicrobial activity against *Escherichia coli*, *Staphylococcus aureus*, *Pseudomonas aeruginosa* and *Klebsiella pneumonia* [64]. In diabetic patients, reduction in growth factors and weakened angiogenesis lead to impaired healing. Chitosan hyaluronic acid composite sponge loaded with VEGF containing fibrin nanoparticles was designed, which showed induction of angiogenesis during wound healing [65].

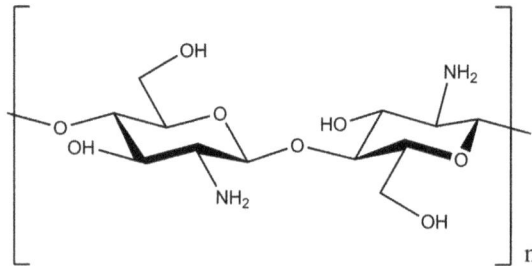

Figure 5. Structure of chitosan having two units; D-glucosamine and N-acetyl-D-glucosamine linked by β (1-4) glycosidic linkages.

To obtain an efficient dressing having control on fluid loss, various chitosan based sponges were designed. A combination of hydroxybutyl chitosan and chitosan was used to develop a composite sponge. Due to superior hydrophilicity, the composite sponge has coagulant properties that helped in controlling the blood loss and also absorbed excessive exudates from the wound. The designed sponges promoted the growth of fibroblast, helped in formation of skin glands and enhance reepithelization with no cytotoxicity [66]. Similarly, a homeostatic composite sponge was developed by blending chitosan, squid ink polysaccharide and calcium chloride. The coagulation ability of the sponge prevented bleeding and maintained homeostasis. Moreover, the sponge promoted healing in scalded rabbits and also controlled infection [67].

A novel asymmetric membrane was prepared by immersion precipitation phase-inversion method and its tissue regeneration properties were evaluated. The top layer contained chitosan based skin surface and interconnected pores designed to limit the bacterial invasion and retain moisture at wound surface [68]. The sponge like sublayer allows high fluid absorption. Microscopic images of healed skin confirmed rapid epithelization and well-organized collagen deposition, suggesting its potential in wound healing. Findings from these studies suggest that characteristic features of chitosan make it an ideal biomaterial for developing smart wound healing products.

6.5   Dextran composite sponges

Dextran is a water soluble, branched polymer composed of glucose subunits joined together by α (1,6) linkages as illustrated in Fig. 6. Dextran is a non-toxic hydrophilic polysaccharide with a well-known biocompatibility proved by various *in-vitro* and *in-vivo* studies. Therefore, it has been extensively explored for biomedical and pharmaceutical applications [69]. Dextran has potential applications as gel in laboratories, and also in medical applications, such as an anti-thrombotic agent, lubricant for eyes and agent for increasing blood sugar [70]. It has been widely explored as wound healing material due to antiseptic and hemostatic characteristics of dextran. Studies showed that dextran hydrogels promoted the angiogenesis and growth of fibroblasts, thus accelerated the skin regeneration in mice with burn wounds [71]. The chemically reactive hydroxyl groups allowed dextran to be tailored with desirable functionalities, and to develop composites with specific characteristics. The chemical and physical modification of dextran has been extensively studied for developing dressing material for efficient wound management.

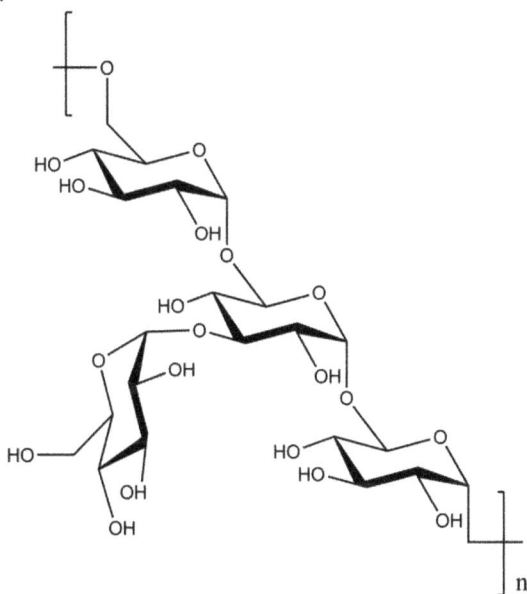

Figure 6. structure of dextran having glucose subunits joined together by a (1,6) linkages.

In one such study, a composite sponge of dextran, collagen and flufenamic acid was developed. The sponges showed good ability for fluid uptake quantified by a high swelling ratio. The treatment of experimentally induced burns on animals with these sponges accelerated the wound healing process and promoted faster regeneration of the affected epithelial tissues compared to the control group. The results generated by the complex sponge characterization indicated that these formulations could be successfully used for dressing applications of burns [72]. The water absorption and wet adhesive properties of dextran make it an efficient homeostatic material. This homeostatic performance of dextran was further improved by fabricating aldehyde-modified, multifunctional poly dextran aldehyde (PDA) sponges through lyophilization. The hemostatic efficacy of theses sponges was investigated both *in-vitro* and *in-vivo*. Moreover, the coagulation mechanisms and biocompatibility was also investigated. The results demonstrated that these sponges could rapidly seal the wound, induce RBCs and platelets aggregation, highly concentrate the coagulation factors, promote the blood coagulation, and achieve remarkable massive hemorrhage control in the ear vein, femoral artery and liver injuries of rats and rabbit models. Additionally, these sponges exhibited low cytotoxicity, no skin irritation, and commendable degradation behavior, therefore, can be used as a homeostatic dressing for wounds with uncontrolled hemorrhage [73].

The hemostatic porous sponge was fabricated by using cationized dextrans and HA via self-foaming process. Due to higher cationic charges, composite sponge displayed higher porosity, better swelling ratios and good blood compatibility [74]. These novel polysaccharide based sponges are worthy for further investigation and may offer a new strategy for designing high-performance wound healing biomaterials.

6.6   Other composite sponges of the polysaccharides

Some other polysaccharides were also investigated for healing activity such as the composite sponge of Konjac glucomannan and silver nanoparticle. The designed composite had good mechanical properties, water holding capacity and cytocompatibility. The presence of silver nanoparticles further improved the healing properties of the composite sponge by introducing antimicrobial activity [75]. In another study, Konjac glucomannan and silk fibroin was physically crosslinked to fabricate a protein polysaccharide porous sponge with excellent biocompatibility. Increasing the concentration of Konjac glucomann significantly enhanced the water absorption and retaining properties of the sponge. Furthermore, the sponge favours the cellular adhesion and proliferation and thus showed potential application in wound dressing [76].

Bletilla striata is widely used in traditional Chinese medicines as a homeostatic agent. Recently, a composite sponge was prepared by combining the *B. striata* and graphene oxide via solution mixing protocol followed by freeze drying. The resulting composite sponge was highly porous with enhanced blood absorption property. This composite is safe and low cost option for developing hemostatic agent with great wound healing potential [77]. Sponges are frequently used as a homeostatic material for bleeding wounds. A composite sponge was prepared by synergizing the hemostatic property of chitosan, alginate and *B. striata* that shortens the homeostatic time and stops bleeding when tested through rabbit ear artery experiments. The *in vitro* cytotoxic analysis further confirmed the biocompatibility of the composite sponge [78]. It is expected that various other polysaccharide will find application in wound healing applications.

Conclusion

Natural polysaccharides are gaining considerable attention amongst researchers for wound healing applications due to their significant biomedical properties like biocompatibilty, nontoxicity, biodegradability, antimicrobial and hemostatic property. The capability of these polymers to regulate the different stages of healing have been exploited to design various wound healing products. This chapter covered the brief evolution of wound care therapies ranging from traditional treatments to bioactive dressing systems. Moreover, the wound healing mechanism was also discussed. This chapter presents the literature concerning polysaccharide based advanced dressings with special emphasis on the composite sponges, polymeric blends and their applications for various cutaneous wounds. However, a much better understanding of the healing mechanisms of these composite sponges is required to cope with the challenges of developing an ideal dressing system.

References

[1]　E. Mclafferty, C. Hendry, A. Farley, The integumentary system: anatomy,physiology and function of skin, Nurs. Stand. 27 (2012) 35. https://doi.org/10.7748/ns2012.10.27.7.35.c9358

[2]　L. Watret, R. White, Surgical wound management: the role of dressings. Nurs. Stand. 44 (2001) 59-69.

[3]　B. M. Delavary, W. M. van der Veer, M. van Egmond, F. B. Niessen, R. H. Beelen, Macrophages in skin injury and repair, Immunobiology. 216 (2011) 753-762. https://doi.org/10.1016/j.imbio.2011.01.001

[4]　P. Martin, Wound healing-aiming for perfect skin regeneration, Science. 276 (1997) 75-81. https://doi.org/10.1126/science.276.5309.75

[5]　R. A. Clark, The molecular and cellular biology of wound repair, Ed Springer Science & Business Media, 2013.

[6]   B. Behm, P. Babilas, M. Landthaler, S. Schreml, Cytokines, chemokines and
        growth factors in wound healing, J. Eur. Acad. Dermatol. Venereol. 26 (2012)
        812-820. https://doi.org/10.1111/j.1468-3083.2011.04415.x

[7]   S. Barrientos, O. Stojadinovic, M.S. Golinko, H. Brem, M. Tomic-Canic, Growth
        factors and cytokines in wound healing, Wound Repair Regen.16 (2008) 585-601.
        https://doi.org/10.1111/j.1524-475X.2008.00410.x

[8]   T. Velnar, T. Bailey, V. Smrkolj, The wound healing process: an overview of the
        cellular and molecular mechanisms, J. Int. Med. Res. 37 (2009) 1528-1542.
        https://doi.org/10.1177/147323000903700531

[9]   N. B. Menke, K. R. Ward, T. M. Witten, D. G. Bonchev, R. F. Diegelmann,
        Impaired wound healing, Clin. Dermatol. 25 (2007) 19-25.
        https://doi.org/10.1016/j.clindermatol.2006.12.005

[10]   G. C. Gurtner, S.Werner, Y. Barrandon, M. T. Longaker, Wound repair and
        regeneration, Nature. 453 (2008) 314. https://doi.org/10.1159/000339613

[11]   T. J. Koh, L. A. DiPietro, Inflammation and wound healing: the role of the
        macrophage, Expert Rev. Mol. Med. 13 (2011).
        https://doi.org/10.1017/S1462399411001943

[12]   A. Bishop, S. Witts, T. Martin, The role of nutrition in successful wound healin, J.
        Community Nurs. 32 (2018) 44-50.

[13]   S. Uddin, A. Bayat, Non-invasive objective devices for monitoring the
        inflammatory, proliferative and remodelling phases of cutaneous wound healing
        and skin scarring, Exp dermatol. 25, (2016) 579-585.
        https://doi.org/10.1111/exd.13027

[14]   M. P. Caley, V. L. Martins, E. A. O'Toole, Metalloproteinases and wound healing,
        Adv wound care. 4 (2015) 225-234.
        https://dx.doi.org/10.1089%2Fwound.2014.0581

[15]   O. Moreno-Arotzena, J. Meier, C. del Amo, J. García-Aznar, Characterization of
        fibrin and collagen gels for engineering wound healing models, Materials. 8
        (2015) 1636-1651. https://doi.org/10.3390/ma8041636

[16]   L. K. Branski, D. N. Herndon, R. E. Barrow, A brief history of acute burn care
        management, In Total Burn Care. (2018) 1-7. http://dx.doi.org/10.1016/B978-0-
        323-47661-4.00001-0

[17]   C. Daunton, S. Kothari, L. Smith, D. Steele, A history of materials and practices for wound management, Wound Practice & Research: Journal of the Australian Wound Management Association. 20 (2012) 174.

[18]   S. Dhivya, V.V. Padma, E. Santhini, Wound dressings–a review, BioMedicine. 5 (2015) 22. https://dx.doi.org/10.7603%2Fs40681-015-0022-9

[19]   A.P. Kornblatt, V. Nicoletti, A. Travaglia, The neglected role of copper ions in wound healing, J. Inorg. Biochem.161 (2016) 1-8. https://doi.org/10.1016/j.jinorgbio.2016.02.012

[20]   K.I. Mohr, History of antibiotics research, In How to Overcome the Antibiotic Crisis Springer, Cham. (2016) 237-272. https://doi.org/10.1007/82_2016_499

[21]   R. F. Pereira, P. J. Bartolo, Traditional therapies for skin wound healing, Adv. Wound Care. 5 (2016) 208-229. https://dx.doi.org/10.1089%2Fwound.2013.0506

[22]   D. Simões, S. P. Miguel, M. P Ribeiro, P. Coutinho, A. G. Mendonça, I. J. Correia, Recent advances on antimicrobial wound dressing: A review, Eur J. Pharm. Biopharm.127 (2018) 130-141. https://doi.org/10.1016/j.ejpb.2018.02.022

[23]   G. D. Winter, Formation of the scab and the rate of epithelization of superficial wounds in the skin of the young domestic pig, Nature. 193 (1962) 293. https://doi.org/10.1038/193293a0

[24]   M. Mir, M. N. Ali, A. Barakullah, A. Gulzar, M. Arshad, S. Fatima, M. Asad, Synthetic polymeric biomaterials for wound healing: a review, Prog Biomater. 7 (2018) 1-21. https://doi.org/10.1007/s40204-018-0083-4

[25]   G. Dabiri, E. Damstetter, T. Phillips, Choosing a wound dressing based on common wound characteristics, Adv. wound care. 5 (2016) 32-41. https://doi.org/10.1089/wound.2014.0586

[26]   G. D. Mogoşanu, A. M. Grumezescu, Natural and synthetic polymers for wounds and burns dressing, Int. J. Pharm. 463 (2014) 127-136. https://doi.org/10.1016/j.ijpharm.2013.12.015

[27]   R. Zafar, K. M. Zia, S. Tabasum, F. Jabeen, A. Noreen, M. Zuber, Polysaccharide based bionanocomposites, properties and applications: A review, Inter. J. Biol. Macromol. 92 (2016) 1012-1024. https://doi.org/10.1016/j.ijbiomac.2016.07.102

[28]   M. Iwamoto, M. Kurachi, T. Nakashima, D. Kim, K.Yamaguchi, T. Oda, T. Muramatsu,. Structure–activity relationship of alginate oligosaccharides in the induction of cytokine production from RAW264, 7 cells. FEBS Letters. 579 (2005) 4423-4429. https://doi.org/10.1016/j.febslet.2005.07.007

[29] C. A. Ryan, E. E. Farmer. Oligosaccharide signals in plants: A current assessment, Annu Rev Plant Physiol. Plant Mol. Biol. 42 (1991) 651-674. https://doi.org/10.1146/annurev.pp.42.060191.003251

[30] R. M. Bloebaum, J.A. Grant, S. Sur, Immunomodulation: The future of allergy and asthma treatment, Curr Opin Allergy Clin Immunol. 4 (2004) 63-67.

[31] L. Sun, Y. Zhao, The biological role of dectin-1 in immune response, Int. Rev. Immunol. 26 (2007) 349-364. https://doi.org/10.1080/08830180701690793

[32] P. R. Martins, A. M. V. Soares, A. V. S. P. Domeneghini, M. A. Golim, R. Kaneno, Agaricusbrasiliensis polysaccharides stimulate human monocytes to capture Candida albicans, express toll-like receptors 2 and 4, and produce pro-inflammatory cytokines, J. Venom Anim. Toxins Incl. Trop. Dis. 23 (2017) 17. https://doi.org/10.1186%2Fs40409-017-0102-2

[33] C. A. Dinarello, Biologic basis for interleukin-1 in disease, Blood. 87 (1996) 2095-2147

[34] S. Echeverry, X. Q. Shi, A. Haw, H. Liu, Z. W. Zhang, J. Zhang, Transforming growth factor-beta1 impairs neuropathic pain through pleiotropic effects, Mol. Pain. 5 (2009) 16. https://doi.org/10.1186%2F1744-8069-5-16

[35] B. Zhao, X. Zhang, W. Han, J. Cheng, Y. Qin, Wound healing effect of an Astragalus membranaceus polysaccharide and its mechanism, Mol. Med. Rep. 15 (2017) 4077-4083. https://doi.org/10.3892%2Fmmr.2017.6488

[36] C. H. Wang, S. J. Chang, Y. S. Tzeng, Y. J., Shih, C. Adrienne, S. G. Chen, J. H.Cherng, Enhanced wound healing performance of a phyto polysaccharide enriched dressing–a preclinical small and large animal study, Int. wound J. 14, (2017) 1359-1369. https://doi.org/10.1111/iwj.12813

[37] B. A. Aderibigbe, B. Buyana, Alginate in Wound Dressings, Pharmaceutics. 10 (2018) 42. https://doi.org/10.3390%2Fpharmaceutics10020042

[38] J. Melrose, Glycosaminoglycans in wound healing. Bone and Tissue Regeneration Insights. 7 (2016) 29-50. http://dx.doi.org/10.1155/2015/834893

[39] P. V. Peplow, Glycosaminoglycan: A candidate to stimulate the repair of chronic wounds, Thromb. Haemost. 94. (2005) 4-16. https://doi.org/10.1160/TH04-12-0812

[40] M. Iwamoto, M. Kurachi, T. Nakashima, D. Kim, K. Yamaguchi, T. Oda, T. Uramatsu, Structure–activity relationship of alginate oligosaccharides in the

induction of cytokine production from RAW264. 7 cells, FEBS letters. 579 (2005) 4423-4429. https://doi.org/10.1016/j.febslet.2005.07.007

[41]   X. Shi, Q. Fang, M. Ding, J. Wu, F. Ye, Z. Lv, J. Jin, Microspheres of carboxymethyl chitosan, sodium alginate and collagen for a novel hemostatic in vitro study, J. Biomater. Appl. 30 (2016) 1092-1102. https://doi.org/10.1177/0885328215618354

[42]   D. H. Roh, S. Y. Kang, J. Y. Kim, Y. B. Kwon, H. Y., Kweon, K. G. Lee, J. H. Lee, Wound healing effect of silk fibroin/alginate-blended sponge in full thickness skin defect of rat, J. Mater. Sci.: Mater Med. 17 (2006) 547-552. https://doi.org/10.1007/s10856-006-8938-y

[43]   R.J. Moon, A. Martini, J. Nairn, J. Simonsen, J.J. Young, Cellulose nanomaterials review: structure, properties and nanocomposites, Chem. Soc. Rev. 40 (2011) 3941-3994. https://doi.org/10.1039/c0cs00108b

[44]   T.R. Stumpf, X. Yang, J. Zhang, X. Cao, In situ and ex situ modifications of BC for applications in tissue engineering, Mater. Sci.Eng: C. 82 (2018) 372-383. https://doi.org/10.1016/j.msec.2016.11.121

[45]   A. Khalid, R. Khan, M. Ul-Islam, T. Khan, F. Wahid, BC-zinc oxide nanocomposites as a novel dressing system for burn wounds, Carbohydr. Polym.164 (2017) 214-221. http://doi.org/10.1016/j.carbpol.2017.01.061

[46]   S. Ye, L. Jiang, J. Wu, C. Su, C. Huang, X. Liu, W. Shao, Flexible amoxicillin-grafted BC sponges for wound dressing: in vitro and in vivo evaluation, ACS Appl. Mater. Interfaces. 10 (2018) 5862-5870. https://doi.org/10.1021/acsami.7b16680

[47]   S. Gustaite, J. Kazlauske, J. Obokalonov, S. Perni, V. Dutschk, J. Liesiene, P. Prokopovich, Characterization of cellulose based sponges for wound dressings, Colloids Surf A: Physicochem. Eng. Asp. 480 (2015) 336-342. https://doi.org/10.1016/j.colsurfa.2014.08.022

[48]   H. O. Barud, H. D. Barud, M. Cavicchioli, T. S. do Amaral, O. B. de Oliveira Junior, D. M. Santos, P. F. de Oliveira, Preparation and characterization of a BC/silk fibroin sponge scaffold for tissue regeneration. Carbohydr. Polym.128 (2015) 41-51. https://doi.org/10.1016/j.carbpol.2015.04.007

[49]   S. Kirdponpattara, M. Phisalaphong, S. Kongruang, Gelatin-BC composite sponges thermally cross-linked with glucose for tissue engineering applications,

Carbohydr. Polym. 177 (2017) 361-368.
http://doi.org/10.1016/j.carbpol.2017.08.094

[50] M. Litwiniuk, A. Krejner, M. S., Speyrer, A. R. Gauto, T. Grzela, Hyaluronic acid
      in inflammation and tissue regeneration, Wounds. 28 (2016) 78-88.

[51] R. D. Price, S. Myers, I. M. Leigh, H. A. Navsaria, The role of hyaluronic acid in
      wound healing, Am. J. Clin. Derm. 6 (2005) 393-402.
      https://doi.org/10.2165/00128071-200506060-00006

[52] S. Kondo, Y. Kuroyanagi, Development of a wound dressing composed of
      hyaluronic acid and collagen sponge with epidermal growth factor, J. Biomat. Sci.
      Polym. Ed. 23(2012) 629-643. https://doi.org/10.1163/092050611X555687

[53] Y. Matsumoto, Y. Kuroyanagi, Development of a wound dressing composed of
      hyaluronic acid sponge containing arginine and epidermal growth factor, J.
      Biomat. Sci. Polym. Ed. 21 (2010) 715-726.
      https://doi.org/10.1163/156856209X435844

[54] C. Longinotti, The use of hyaluronic acid based dressings to treat burns: A review,
      Burns trauma. 2 (2014) 162. https://doi.org/10.4103%2F2321-3868.142398

[55] Y. K.Lin, D. C. Liu, Studies of novel hyaluronic acid-collagen sponge materials
      composed of two different species of type I collagen, J. Biomat. Appl. 21 (2007)
      265-281. https://doi.org/10.1177/0885328206063502

[56] M. Mahedia, N. Shah, B. Amirlak, Clinical evaluation of hyaluronic acid sponge
      with zinc versus placebo for scar reduction after breast surgery, Plast. Reconstr.
      Surg. Glob. 4 (2016). https://doi.org/10.1097/GOX.0000000000000747

[57] C. J. Doillon, F. H. Silver, R. A. Berg, Fibroblast growth on a porous collagen
      sponge containing hyaluronic acid and fibronectin, Biomaterials. 8 (1987) 195-
      200. https://doi.org/10.1016/0142-9612(87)90063-9

[58] M. Rinaudo, Chitin and chitosan: properties and applications, Prog. Polym. Sci. 31
      (2006) 603-632. https://doi.org/10.1016/j.progpolymsci.2006.06.001

[59] V.Patrulea, V. Ostafe, G. Borchard, O. Jordan, Chitosan as a starting material for
      wound healing applications, Eur J. Pharm. Biopharm. 97(2015) 417-426.
      https://doi.org/10.1016/j.ejpb.2015.08.004

[60] D. Mei, Z. XiuLing, X. Xu, Y. K. Xiang, Y. L. Xing, G. Gang, L. Feng, Z. Xia, Q.
      W. Yu, Q. Zhiyong, Chitosan-alginate sponge: preparation and application in
      curcumin delivery for dermal wound healing in rat, J. Biomed. Biotechnol. (2009)
      8. http://doi.org/10.1155/2009/595126

[61]   J. Wu, C. Su, L. Jiang, S. Ye, X. Liu, W. Shao, Green and Facile Preparation of
       Chitosan Sponges as Potential Wound Dressings, ACS Sustain. Chem. Eng. 6
       (2018) 9145-9152. http://pubs.acs.org/doi/abs/10.1021/acssuschemeng.8b01468

[62]   B. Lu, T. Wang, Z. Li, F. Dai, L. Lv, F. Tang, G. Lan, Healing of skin wounds
       with a chitosan–gelatin sponge loaded with tannins and platelet-rich plasma. Int J
       Biol Macromol. 82 (2016) 884-891.
       https://doi.org/10.1016/j.ijbiomac.2015.11.009

[63]   M. Liu, Y. Shen, P. Ao, L. Dai, Z. Liu, C. Zhou, The improvement of hemostatic
       and wound healing property of chitosan by halloysite nanotubes, RSC Advance, 4
       (2014) 23540-23553. https://doi.org/10.1039/C4RA02189D

[64]   B. S. Anisha, R. Biswas, K. P. Chennazhi, R. Jayakumar, Chitosan–hyaluronic
       acid/nano silver composite sponges for drug resistant bacteria infected diabetic
       wounds. Int. J. Biol. Macromol. 62 (2013) 310-320.
       https://doi.org/10.1016/j.ijbiomac.2013.09.011

[65]   A. Mohandas, B. S. Anisha, K. P. Chennazhi, R. Jayakumar, Chitosan–hyaluronic
       acid/VEGF loaded fibrin nanoparticles composite sponges for enhancing
       angiogenesis in wounds, Colloid Surface B. 127 (2015) 105-113.
       https://doi.org/10.1016/j.colsurfb.2015.01.024

[66]   S. Hu, S. Bi, D. Yan, Z. Zhou, G. Sun, X. Cheng, X. Chen, Preparation of
       composite hydroxybutyl chitosan sponge and its role in promoting wound healing,
       Carbohydr. Polym. 184 (2018) 154-163.
       https://doi.org/10.1016/j.carbpol.2017.12.033

[67]   N. Huang, J. Lin, S. Li, Y. Deng, S. Kong, P. Hong, Z. Hu, Preparation and
       evaluation of squid ink polysaccharide-chitosan as a wound-healing sponge,
       Mater. Sci Eng: C. 82 (2018) 354-362. https://doi.org/10.1016/j.msec.2017.08.068

[68]   F. L. Mi, S. S. Shyu, Y. B. Wu, S. T. Lee, J. Y. Shyong, R. N. Huang, Fabrication
       and characterization of a sponge-like asymmetric chitosan membrane as a wound
       dressing, Biomaterials. 22 (2001) 165-173. https://doi.org/10.1016/s0142-
       9612(00)00167-8

[69]   R. I. Malini, J. Lesage, C. Toncelli, G. Fortunato, R. M. Rossi, F. Spano,
       Crosslinking dextran electrospun nanofibers via borate chemistry: Proof of
       concept for wound patches, Eur. Polym J. 110 (2019) 276-282.
       https://doi.org/10.1016/j.eurpolymj.2018.11.017

[70]   M. Ghica, M. Albu Kaya, C. E. Dinu-Pîrvu, D. Lupuleasa, D. Udeanu, Development, optimization and in vitro/in vivo characterization of collagen-dextran spongious wound dressings loaded with flufenamic acid, Molecules. 22 (2017) 1552. https://doi.org/10.3390/molecules22091552

[71]   G. Sun, X. Zhang, Y. I. Shen, R. Sebastian, L. E. Dickinson, K. Fox-Talbot, S. Gerecht, Dextran hydrogel scaffolds enhance angiogenic responses and promote complete skin regeneration during burn wound healing, Proc. Natl. Acad. Sci.108 (2011) 20976-20981. https://doi.org/10.1073/pnas.1115973108

[72]   M. Ghica, M. Albu Kaya, C. E. Dinu-Pîrvu, D. Lupuleasa, D.Udeanu, Development, optimization and in vitro/in vivo characterization of collagen-dextran spongious wound dressings loaded with flufenamic acid, Molecules. 22, (2017) 1552. https://doi.org/10.3390/molecules22091552

[73]   C. Liu, X. Liu, C. Liu, N. Wang, H. Chen, W. Yao, W. Qiao, A highly efficient, in situ wet-adhesive dextran derivative sponge for rapid hemostasis, Biomaterials. 205 (2019) 23-37. https://doi.org/10.1016/j.biomaterials.2019.03.016

[74]   J.Y. Liu, Y. Li, G. Hu, G. Cheng, E. Ye, C. Shen, F.J. Xu, Hemostatic porous sponges of cross-linked hyaluronic acid/cationized dextran by one self-foaming process, Mat. Sci Eng: C. 83 (2018) 160-168. https://doi.org/10.1016/j.msec.2017.10.007

[75]   H. Chen, G. Lan, L. Ran, Y. Xiao, K. Yu, B. Lu, F. Lu, A novel wound dressing based on a Konjac glucomannan/silver nanoparticle composite sponge effectively kills bacteria and accelerates wound healing, Carbohydr. Polym. 183 (2018) 70-80. https://doi.org/10.1016/j.carbpol.2017.11.029

[76]   Y. Feng, X. Li, Q. Zhang, S. Yan, Y. Guo, M. Li, R. You, Mechanically robust and flexible silk protein/polysaccharide composite sponges for wound dressing, Carbohydr. Polym. 216 (2019) 17-24. https://doi.org/10.1016/j.carbpol.2019.04.008

[77]   J. Chen, l. Lv, Y. Li, X. Ren, H. Luo, Y. Gao, X. Li, Preparation and evaluation of Bletillastriata polysaccharide/graphene oxide composite hemostatic sponge, Int. J. Biol Macromol.130 (2019) 827-835. https://doi.org/10.1016/j.carbpol.2017.06.112

[78]   C. Wang, W. Luo, P. Li, S. Li, Z. Yang, Z. Hu, N. Ao, Preparation and evaluation of chitosan/alginate porous microspheres/Bletillastriata polysaccharide composite hemostatic sponges, Carbohydr. Polym. 174 (2017) 432-442. https://doi.org/10.1016/j.carbpol.2017.06.112

Advanced Applications of Polysaccharides and their Composites　　　Materials Research Forum LLC
Materials Research Foundations **73** (2020) 27-64　　　　　　　　https://doi.org/10.21741/9781644900772-2

Chapter 2

Polysaccharides for Drug Delivery

Kamla Pathak*[1], Rishabha Malviya[2]

[1]Faculty of Pharmacy, Uttar Pradesh University of Medical Sciences, Etawah, 206130, Uttar Pradesh, India

[2]Department of Pharmacy, School of Medical and Allied Sciences, Galgotias University, Greater Noida, Uttar Pradesh, India

*kamlapathak5@gmail.com

Abstract

Polysaccharides are complex versatile biomaterial, which are generally biocompatible, biodegradable and non-toxic in nature. Since their affirmed position as valuable pharmaceutical excipients for development of a plethora of dosage forms, these have been extensively explored as carriers for drug delivery. The chapter elaborates the applications of polysaccharides and their derivatives for sustained/targeted delivery of various categories of therapeutic agents. Patents have also been listed to emphasize commercial sustainability of the polysaccharide based drug delivery systems. Furthermore, newer applications namely polysaccharide anchored liposomes, auto-associated amphiphilic polysaccharides, electrospun polysaccharides, polyelectrolyte complexes, polysaccharide based systems namely aerogel, nanocomposites, nanogels, quantum dots, nanoparticles and micelles have also been described. The drug release kinetics from the polysaccharidic systems has been detailed.

Keywords

Polysaccharides, Drug Carriers, Advancements, Drug Release Kinetics

Contents

1.   Introduction

Polysaccharides are promising biodegradable polymeric materials used in pharmaceutical, food, cosmetic and biotechnology industry. Conventionally polysaccharides have been used as excipient in pharmaceutical formulations, but in novel dosage forms they frequently perform significant multifunctional roles namely drug release modulators, bioavailability modifiers, stability enhancers and ensure manufacturing amenability [1]. While various synthetic polymers are available to a pharmaceutical formulator; the use of polysaccharides (obtained naturally) offer an attractive option because of their abundance, amenability to synthetic (chemical) alterations, devoid of toxicity, biodegradability and by and large being biocompatible. Furthermore, sustainable cultivation and harvest techniques can assure continual supply of the pharmaceutical raw material. A variety of pharmaceutical dosage forms like matrix systems, buccal films, microparticles, nanoparticles, ophthalmic solutions/suspensions, implants have been formulated and successfully commercialized based on use of polysaccharides [2, 3].

Advanced Applications of Polysaccharides and their Composites   Materials Research Forum LLC
Materials Research Foundations **73** (2020) 27-64   https://doi.org/10.21741/9781644900772-2

Chemically, polysaccharides are formed by the glycosidic linkages of monosaccharides [4] and based on the nature of the monosaccharide; the polysaccharides are divided into linear and branched chains [5]. Schematic diagram of polysaccharide classification is shown as Figure 1. Polysaccharides contain various groups on their molecular chains that can be subjected to modification to give different types of polysaccharide derivatives.

Gums and mucilages are hydrocolloids of plants that contain galactose, methyl pentose, and sugars connected through glycosidic linkages to uronic acid residues. Natural gums are polysaccharidic molecules consisting of multiple sugar units linked to yield macromolecules. Acacia, guar gum and tragacanth are few examples of gums used in the pharmaceutical and biotechnology industry. Generally, they are complex carbohydrate polymers which consist of hydroxy-proline rich proteins, resins, and other components. The gums can be soluble, partially soluble or insoluble in nature [4]. Mucilages are metabolic products of plants and, form slimy masses when placed in water. They contain sulfuric acid esters which are composite polysaccharides. Both natural gum and mucilage have the ability to form three dimensional molecular networks. The gelling ability is based on molecular structure, temperature, pH, ionic strength and concentration. The constraints associated with polysaccharides include batch to batch variation, microbial load, pH dependent solubility, uncontrolled hydration rate and viscosity diminution on storage [5].

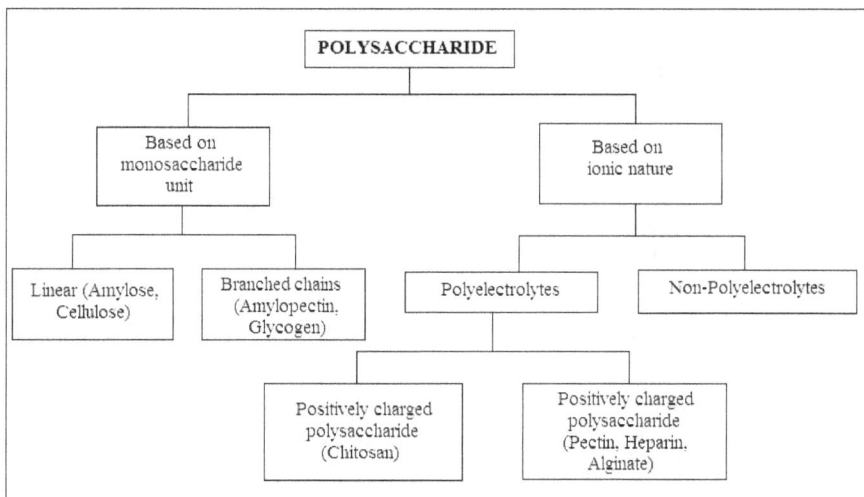

Figure 1. Classification of polysaccharides.

2. Applications

Polysaccharides are extensively used raw material(s) for designing conventional and new delivery systems. Polysaccharides included in drug delivery systems are intended to multitask and apparently control the rate or extent of drug release directly or indirectly. Generally speaking they play a significant role in formulating inexpensive as they essentially form an excipient that is available locally [6]. Table 1 compiles a cross section of the applications of polysaccharides in drug delivery. Few references in each application have been exemplified in the tabulation.

Table 1. Polysaccharides based various drug delivery systems.

Type of delivery	Role of polysaccharides in the system	Reference
Colon drug delivery	In the colon targeting delivery, when concentration of polysaccharide is increase so it enhances the disintegration time and helps in the sustaining and targeting the release of drug in lower intestine. In the presence colonic enzyme, the polysaccharides are biodegraded and ensuring the drug release completely which further helps in the absorption. Example; tamarind seed polysaccharide, guar gum, pectin, chitosan etc.	[1, 2]
Buccal drug delivery	The role of bioadhesive polysaccharide used in buccal drug delivery is to retain a formulation. They are typically hydrophilic macro-molecules containing numerous hydrogen bonding groups. Higher drug release is obtained when the polysaccharide used in the formulation have good mucoadhesive properties. Example; Chitosan, tamarind seed polysaccharide.	[3, 4]
Ophthalmic drug delivery	Polysaccharides prolong the precorneal residence time on the cornea due to their high viscosity and mucoadhesive property. Tamarind seed gum has film forming ability with high tensile strength and flexibility. Polysaccharides are found to be good for the formulation of ophthalmic systems. Example; cellulose, chitin, tamarind gum. HPMC.	[3, 5, 6]
Vaginal drug delivery	Mucoadhesive polymer systems like chitosan, cellulose derivatives, hyaluronic acid derivatives, pectin, tragacanth, starch, carrageenan, sodium alginate and gelatin are capable of delivering the active agent for an extended period at a predictable rate to the vagina. Use of polysaccharide improves therapeutic effect of drug.	[7-9]
Transdermal drug delivery	Chitosan, sodium alginate and tamarind seed polysaccharide are used for the preparation of transdermal formulations. These help in providing controlled release and facilitate permeation across skin	[5,10]
Nasal drug delivery	Polysaccharide provides intimate contact between a dosage form and the absorbing tissue, which may result in high concentration in a local area and hence high drug flux through the absorbing tissue. Polysaccharides are used for the preparation of nasal formulations. Example; chitosan and its derivatives, tamarind gum, sodium alginate.	[11, 12]

3.   Advancements

3.1   Polysaccharide anchored liposomes

In order to develop polysaccharide anchored liposomes, it seems imperative to understand the mechanism of interaction of the given polysaccharide with the liposome bilayer and the impact of the anchoring on the bilayer fluidity, permeability and integrity [19]. For the system to be site specific for targeted delivery, the selectivity and affinity of the polysaccharide towards the ligand(s) is an important prerequisite. Various natural or hydrophobized polysaccharides have been linked on liposomal surface by any of the following methods: adsorption method, polysaccharide induced aggregation or fusion; spacer activated covalent anchoring, and covalent anchoring with sensors [20]. Primarily, the polysaccharides pullulan, dextran, chitosan, amylopectin and mannan are used to develop polysaccharide anchored liposomes. The polysaccharide anchored liposomes also have potential to be employed as carrier platforms for anchoring site-specific sensing molecules like monoclonal antibodies against the tumor surface antigens.

Sunamoto *et al.* explored the interaction of simple polysaccharides with the liposomal membrane. Polysaccharides like chitosan, dextran, mannan, pullulan and amylopectin, were strongly adhered on to the surface of liposome predominantly by hydrophobic bonding followed by clustering and amalgamation of the liposomes [21]. However, in certain conditions the adsorption of polysaccharide molecules on to the liposomal surface is postulated to occur by diffusion-controlled mechanism involving the constitutive liposomal components and the polysachharidic coat. Fluorescence depolarization technique confirmed the lateral diffusion and subsequently interlocking of the adsorbed polysaccharidic molecules within the liposomal bilayer. Adsorption based polysaccharide anchoring is considered to be thermodynamically unstable and pharmaceutical acceptance is questionable owing to the following reasons (i) the polysaccharide molecules adsorbed on the liposomal surface may get readily desorbed/dislodged on mechanical stirring or dilution; (ii) the adsorbed polysaccharide may coagulate leading to destabilization of the liposome bilayer, and (iii) stoichiometry is frequently not reproducible. To overcome these hurdles, Sunamoto and Iwamoto used partially hydrophobized polysaccharides in order accomplish coating of liposomes. The palmitoylated polysaccharides were reacted with the lipidic components of the bilayer to form a polysaccharide based artificial wall on the liposome [22].

Later, the structural stability of polysaccharide decorated liposomes was assessed by using both lipophilic and hydrophilic markers. The radiolabelled *O*-palmitoyl amylopectin decorated liposomes with coenzyme Q10 (^{14}C) in the lipid bilayer and inulin (^{3}H) in the internal aqueous core exhibited stability both in the biological specimens as demonstrated by radioisotope analysis. Additionally, enhanced stability and bilayer intactness of the lipid membranes decorated with polysaccharidic derivatives containing hydrophobic groups (cholesteroyl and palmitoyl group) was also evidenced [23]. The researchers of the same team reported formulation of immunopolysaccharide conjugate anchored liposomes for site specific targeted delivery and claimed *in vitro* stability of the developed system in human serum [24].

Elferink *et al.* evaluated the stability of the proteoliposomes prepared from the phospholipids obtained from *E. coli* and anchored them with hydrophobized dextran. The coating not only conferred stability on the liposomes but also stabilized the proteoliposomes in biological environment [25]. Deol *et al.* developed long circulating polysaccharide anchored liposomes for localization in the lungs. The targeting ability and prolonged life of the liposomes was demonstrated via *in vitro* experiments followed by *in vivo* study in mice. The anchored liposomes were assessed for their effectiveness in chemotherapy of tuberculosis. The stealth (pegylated) liposomal surface was modified by anchoring it with O-stearoyl-amylopectin that enhanced its affinity and consequently resulted in its higher accumulation in lungs than in the organs of reticulo-endothelium system in both normal and tubercular mice. The encapsulated antitubercular drugs rifampicin and isoniazid in the stealth liposomes demonstrated diminutive toxicity in mice [26].

Among the various implication of using polysaccharide anchored liposomes, the site-specific targeting of the vesicle due to its bio-/mucoadhesive properties is of great importance. Takeuchi *et al.* investigated polymer anchored mucoadhesive multilamellar liposomes composed of diacetyl phosphate and dipalmitoyl phosphatidylcholine and anchored them with three types of polymers: poly (acrylic acid) bearing cholesterol, polyvinyl alcohol and chitosan. The chitosan decorated liposomes demonstrated strongest bioadhesion among the decorated liposomes tested while the non-decorated liposomes demonstrated insignificant to complete non-adhesion in rat intestine. The mucoadhesive property of chitosan decorated liposomes was later utilized to formulate oral drug delivery system(s) of the weakly absorbed drugs [27]. In another research, Takeuchi *et al.* explored the mucoadhesive property of chitosan decorated liposomes to enhance the absorption of insulin administered orally in male Wistar rats. Prolonged hypoglycemic response (up to 12 h) was observed on oral administration of insulin loaded chitosan decorated liposomes. The sustained pharmacological effect was attributed to the mucoadhesive property of the delivery system that prolonged the contact time thereby increasing insulin absorption. These reports affirm the ability of polysaccharide anchored liposomes as mucoadhesive drug delivery systems for poorly adsorbed drugs and macromolecules [28].

N-palmitoyl chitosan, a chitosan derivative, is a novel polymer utilized to anchor liposomes and has been explored to stabilize liposomes containing anti-tumor agent. Liang *et al.* prepared N-palmitoyl chitosan anchored docetaxel loaded liposomes with the aim to and improve the antitumor efficacy and reduce its toxic effects. The anchored liposomes effectively increased the stability of docetaxel *in vivo* in comparison to the plain liposomes [29]. In another report Wang *et al.* formulated cholesterol succinyl chitosan anchored liposomes by the incubation method and carried out pharmaceutical investigations. The anchored liposomes were spherical with classical shell core structure. Epirubicin was effectively loaded into chitosan anchored liposomes and the drug showed sustained release both in phosphate buffer, pH 7.4 and 1 %v/v aqueous fetal bovine serum *in vitro* [30]. Another chitosan derivative, N-pamitoyl chitosan was used as anchoring material for oleic acid liposomes. Tan et al. selected water-soluble chitosan with varying degrees of acylation. Reduction in the mean vesicle size and zeta potential of the oleic acid chitosan liposomes suggested successful modification of the liposomal surface via chitosan derivative that enhanced the integrity and rigidity of the vesicles [31].

3.2　Auto- associated amphiphilic polysaccharides

Self-assembly of amphiphilic polysaccharides presents a promising system without the need of surfactants and solvents. In water, various polymeric amphiphilic polysaccharides undergo intra-molecular and inter-molecular interactions. These interactions favoured by hydrophobic segments, paved way for the emergence of various drug delivery systems namely micelles, nanoparticles, hydrogels and liposomes. The selection of auto-associated amphiphilic polysaccharides is based on the physicochemical characteristics of the drug to be loaded and the desired administration route. At higher polymeric concentration, the intermolecular association of polymers is initiated by the association of the hydrophobic groups, which results in the viscosity enhancement of the solution. Hence, these polymers are termed as associating polymers and used as viscosity modifiers. At higher polymeric concentration phase separation or gelation may occur. Chemically modified amphiphilic chitosan was obtained by the grafting hydrophobic groups through the N-acylation reaction because the $-NH_2$ group of chitosan is more reactive than -OH group of chitosan. Auto-associated chitosan form micelles, nanoparticles and hydrogels. The more stable micelles are achieved when amphiphilic chitosan was used with linoleic acid and (1-ethyl-3-(3-dimethylaminopropyl)-carbodiimide) [32].

Apart from chitosan other polysaccharides have also been experimented. Grafting of dextran was done with lauryl chains and stable nanoparticles were formulated using amphiphilic dextran which can auto-associate with poly-b-cyclodextrin [33]. Hyaluronic acid was also chemically modified through di oleoyl phosphatidyl ethanolamine in the presence of coupling agent. The product was utilized in the formulation of liposomes [34]. Amphiphilic pullulan act as a thermoresponsive polymer in the formulation of nanoparticles because it is slightly soluble in aqueous medium at 50 to 60 °C temperature without any precipitation. Hence, Bataille *et al.* introduced long alkyl chain and Glinel *et al.* studied perfluoroalkyl chain in pullulan for auto-associative behaviour [35, 36]. The hydrogel nanoparticles containing cholesteryl bearing pullulans can form complexes in aqueous solution with soluble proteins/ enzymes such as insulin, bovine serum albumin, myoglobin, a-chymotrypsin and cytochrome C. Linoleic acid was incorporated in the amylose solution to obtain the grafted amylose so that it could auto associate in the preparation of micelles [37].

Ayame *et al.* explored the interaction between protein and self-assembled nanogels of cationic cholesteryl group bearing pullulans for effective intracellular protein delivery. The self-assembled cationic nanogels interacted with bovine serum albumin to form stable colloidal monodispersed nanoparticles. The assembly of nanogel protein got dissociated post cellular uptake and the protein got released within the cell. The self-assembled amphiphilic polysaccharides also have potential to provide platform for the delivery of protein [38]. Cao *et al.* reported a simple method of developing self-assembled nanoparticles loaded with anti-cancer drug and the targeting moiety attached to it. Xyloglucon was grafted with galactosamine and doxorubicin was used to target the polymeric conjugate to the hepatocytes. The nanoparticles were formed on mixing the polymeric drug with an excess amount of deprotonated doxorubicin in an aqueous medium. This system can be used for tumor therapy [39]. Kwon *et al.* developed self-aggregates of hydrophobically modified glycol chitosan. The self aggregates were prepared by inter- or intramolecular association between glycol chitosan and 5β-cholanic acid (hydrophobic). The critical aggregation concentration of the hydrophobically modified glycol chitosan and the size of self aggregates, both were dependent on the degree of substitution of 5β-cholanic acid [40].

3.3   Electrospun polysaccharides

Electrospinning is basically a physical method to generate nanocomposite(s) based on the extrusion of polymer solution in the absence/presence of the nanomaterial which is dispersed, by a needle on to a receiver plate, in the presence of high voltage. The resultant fibers have a typical core–sheath structure. It is a stepwise process which requires control over the process parameters and optimization of solution. Primarily, water soluble polysaccharides, namely cellulose acetate, starch and chitosan are used for electrospinning. The use of cellulose and chitin for electrospinning is limited because very few solvents can dissolve them [41].

The polysaccharides namely chitin, alginates, cellulose, starch, chitosan, hyaluronic acid and heparin used in various drug delivery applications are interesting candidates for electrospinning. However, the inadequate solubility of polysaccharides such as chitin and cellulose is the limiting factor. Others such as chitosan, alginate and hyaluronic acid have high viscosity which is attributed to their electrical charges and high molecular weight. Due to this, it also causes poor electrospinnability. These barriers may be overcome by selection of appropriate solvent composition and the polymer blend. The viscosity and conductivity of the polymeric solution are key parameters which affect the characteristics of electrospun fibers. The electrospun blended polyvinyl alcohol/ extracellular polymeric substance membrane demonstrated better mechanical properties and tensile strength when compared to polyvinyl alcohol alone and polyvinyl alcohol /chitosan electrospun fibres [42].

A three dimensional malleable scaffold was fabricated by electrospinning fibrous mats. Hyaluronic acid and alginic acid failed to electrospin because they readily dissolve in water. Blending them with water soluble, biocompatible polymer such as polyvinyl alcohol made them amenable to fiber spinning due to their ability to decrease repulsive forces within the charged biopolymer. Due to the lack of chain entanglements, alginate cannot be electrospun by itself and its nanofibers can be obtained electrospinning only when the alginates are blended with water soluble polymer(s) polyethylene oxide or polyvinyl alcohol. Guar gum has potential to improve electrospinnability by modulating rheological properties because it is related to fiber formation when it is connected with synthetic polymers. Due to high viscosity and low solubility chitosan poses many limitations for electrospinning. The nanofibers of chitosan can be made by electrospinning solution of pure chitosan in concentrated acetic acid/trifluoro acetic acid. Electrospinning of aqueous solution of carboxymethyl chitosan is assisted by addition of water-soluble polymers namely polyacrylic acid, poly(ethylene) oxide, polyvinyl alcohol and polyacrylamide. Zhang et al. studied that compatibility properties were retained by using electrospinning techniques when scaffolds were fabricated by poly lactic acid and angelica saccharide [43]. The drug release behavior of the drug from electrospun nanofibres is determined by degradation of carrier polymer and diffusion of drug molecules. The electrospinning techniques may be precisely controlled to modify the drug release kinetics and to control the distribution the drug in the electrospun fibers.

3.4   Polyelectrolyte complexes for nano drug delivery

Polyelectrolyte complex (also termed as self-assembled polyelectrolyte) are the complexes formed due to self-assemblage of the ionic polymer and the plasmid DNA. Ilina *et al.* studied the chitosan based polyelectrolyte complexes and found that these may be they are used as carriers for active moieties, films, non-viral gene vectors etc. The formation of polyelectrolyte complex depends on the degree of ionization of the cationic and anionic polymer, concentration of polymers, their ratio, charge distribution over the polymeric chain, the temperature of reaction medium and the duration of the interaction [44]. Luo *et al.* explored the chitosan based polyelectrolyte complexes formed using polysaccharides such as xanthan gum, cashew gum, alginate, pectin, hyaluronic acid, carageenan, gellan gum, gum arabic, konjac glucomannan, carboxymethyl cellulose and gum kondagogu in the form of nanoparticles for drug delivery applications [45]. Later, Motwani *et al.* developed polyelectrolyte complex nanoparticles for ocular delivery via chitosan sodium alginate in concentration as low as < 0.1%. The investigators observed that the rate of drug release from the nanoparticles was faster than that observed from micro- particles/beads, which can be attributed to nano size, and consequently large surface area to mass ratio of the polyelectrolyte complex [46]. Erbacher *et al.* studied the mechanism of polyelectrolyte complex via neutralization of charge between the cationic polymer and DNA leading to diminution of hydrophilicity [47]. The alginate and chitosan nanoparticles are used as nanocarriers for controlled release of therapeutic proteins because of the ease of polyelectrolyte complexation between alginate and chitosan.

Teng *et al.* formulated complex nanoparticles of soy protein isolate and carboxymethyl chitosan by ionic gelation method. Vitamin D (hydrophobic micronutrient) was efficiently loaded into the polymeric complex based nanoparticles (162 - 243 nm). Their compact structure rendered an encapsulation efficiency of 96.8% and efficient hydrogen bonding capability as evidenced by FTIR spectroscopy. The polymeric complex nanoparticles exhibited cumulative drug release of 42.3% in simulated gastric condition and about 36% release in simulated intestinal fluid. Thus the developed nanoparticles presented a versatile option for controlled delivery of hydrophobic bioactives and nutraceuticals [48]. Wang *et al.* developed the polyelectrolyte macro ion complexes in the form of nanocrystals between chitosan and cellulose nanocrystals. Prepared nanoparticles were anionic and cylindrical in nature. The size of the polyelectrolyte macro ion complexes was dependent on the ratio in which the formulation components were mixed. Furthermore, high concentration of the chitosan produced positively charged spherical particles whereas its lower concentration formed negatively charged non-spherical particles [49].

Du *et al.* prepared the polyelectrolyte complex of carboxymethyl konjac glucomannan chitosan nanoparticles driven by electrostatic interactions for drug delivery of water soluble drugs. The nanoparticles were explored for their potential as a delivery system for bovine serum albumin. The nanoparticles ranged between 50 and 1200 nm, and the zeta potential varied from 15 to 45 mV. The mean particle size increased and the zeta potential decreased on changing the pH value of the medium and increasing the salt concentration. The study suggested utilization of nanoparticulate system as an advanced drug delivery carrier for water soluble drugs [50]. The polyelectrolyte complex of chitosan and alginate is extensively used to devise microcapsules for cell encapsulation and for development of controlled release formulations. The drug release is grossly affected by the magnitude of swelling of the microsystems. In a study by Pasparakis *et al.,* the factors affecting the swelling of calcium alginate–chitosan microbeads were identified as the ability of polyelectrolyte complexation between chitosan and alginate, the initial physical state of the microbeads and the pH of the swelling medium [51].

Hamman *et al.* observed that the chitosan containing polyelectrolyte complexes show good physicochemical properties. These complexes can therefore serve as excipients for designing various delivery systems namely microparticles, nanoparticles, microbeads, sponges, fibers, sponges and matrix tablets [52]. Saether *et al.* developed the polyelectrolyte complexes by using alginate and chitosan. The factors that influenced the particle size and zeta potential were molecular weight and net charge ratio between chitosan and alginate [53]. Bigucci *et al.* claimed that the polyelectrolyte complexes of chitosan and pectin showed pH dependent swelling and hence the drug release was pH dependent. The said polyelectrolyte complexes were proposed for colon-targeting of vancomycin [54]. Tapia *et al.* studied the polyelectrolyte complexes of chitosan with carrageenan and alginate as prolonged release systems for diltiazem chlorhydrate. The researchers concluded that the chitosan–alginate matrix system was superior to the chitosan–carrageenan matrix system for prolonging drug release [55].

Shu *et al.* prepared microbeads of tripolyphosphate/chitosan complex for controlled drug delivery. Sodium alginate (polyanion) was reacted with cationic chitosan on the surface of these microbeads to form polyelectrolyte complex film that had potential for sustainment of drug release [56]. Maciel *et al.* synthesized the chitosan/carboxymethyl cashew gum polyelectrolyte complexes. At low temperature, they showed more decomposition than the original polysaccharides [57]. Masotti *et al.* reported preparation of chitosan/DNA nanospheres for pH-dependent modulated release of DNA [58]. The spherical nanospheres with an average diameter of 38 ± 4 nm with an encapsulation efficiency of 30% were obtained. Sustained DNA release for over 60 hours was achieved at pH 7.4.

3.5   Polysaccharide based aerogel

Aerogels provide value added light weight platform with excellent surface area and porosity appropriate for drug loading. The aerogels not only exhibit high drug adsorption in their matrix but also depict high drug loading capacity. Aerogels are usually produced from wet gels using supercritical drying technique. This technology possesses integral porous texture of the wet material and is devoid of pore collapse phenomenon while drying. Large surface of aerogels are formed because this technique prevents the gel structure. When the liquid is removed from the gel, aerogels are achieved in different size and shapes such as spherical beads, disks like structure etc. Polysaccharide based aerogels have good mechanical properties for artificial heart valve so they are used for the cardiovascular implantable devices. Also used as a tissue engineering substrate because of their physical properties. In wound healing process, they act as barrier for microorganisms and provide moisture on the wound surface [59].

The aerogels have compressive modulus which is dependent on the type of polysaccharides, ambient moisture and oxidation reactions. Starch and alginate are relatively common polysaccharides for drug formulation due to their low toxicity. The combination of aerogels with structural properties and physiological compatibility of polysaccharides have high potential for drug delivery system. Mainly gelatin, agar, cellulose are used for the preparation of aerogels. Hemicellulose, starch and marine polysaccharides may also be used in the preparation of aerogels. They can readily form gel in the presence of water, polysaccharides blend and cross-linking agents. The aerogels prepared with starch as microspheres can be used for nasal, oral and parenteral drug deliveries for chemotherapy of lung cancer and tissue engineering. Aerogels are also developed by the gelation of chitosan and montmorillonite material in the form of microspheres and are extremely superior to other aerogels. In ambient environment, the pectin aerogels are also prepared which have some thermal conductivity.

Polysaccharide based aerogels are designed for mucosal delivery through nasal, buccal, intestinal and vaginal route due to their mucoadhesive properties. When biomaterials are connected with polysaccharides they increase the loading capacity of aerogels, stability of drug and also modify the drug release behaviour. The composite material silica chitosan aerogels showed interesting results for drug delivery. Aerogels are used as drug carrier in stimuli triggered delivery and show alteration in drug release behaviour. De Cicco *et al.* studied that the aerogel with pectin and alginate can be used in the treatment of wounds infections [60]. Quignard *et al.* prepared aerogels by using marine polysaccharides by supercritical drying process using polymers with variety of functional groups: carrageenans (sulfonic group), alginates (-COOH) and chitosan ($-NH_2$ group) as dry materials which disperse to form polymeric hydrogel [61].

Tkalec *et al.* produced ethanol induced gelation aerogels of alginates, pectin, guar gum and xanthan gum. Polysaccharidic aerogels were formed by dissolving the polysaccharide in water, ethanol induced gelation and supercritical drying as the sequential steps. The highlight of the method as claimed by the authors was gelation by ethanol, without the use of cross-linker and the method was devoid of solvent exchange step. The overall production time was significantly reduced and highly porous monolithic systems with large surface area were obtained [62]. Alnaief *et al.* developed the biodegradable nanoporous micro spherical aerogel containing alginate using emulsion technique. Emulsification of sodium alginate solution in the oil resulted in water in oil emulsion. The droplets of the dispersed phase were cross-linked to form the gel particles [63]. Alnaief *et al.* also developed the microspherical aerogels by emulsion based method. A novel method based on *in situ* emulsification was reported to produce biocompatible spherical aerogel microparticles followed by supercritical drying. Mixing the sol with vegetable oil resulted in water in oil emulsion that was followed by supercritical drying of the dispersed phase to produce aerogels. The aerogels, as proposed by the researchers can be used for designing modulated drug delivery systems [64].

Another interesting research on aerogels based on the whey protein made by supercritical drying technique was reported by Betz *et al.* The strong covalent disulphide bonds of the whey protein based hydrogels prevented the collapse of the aerogel structure on supercritical drying. The aerogel showed pH-dependent swelling upon rehydration. The drug loaded whey protein based aerogels showed sustained drug release in the media maintained at pH 1.2 and in the simulated intestinal pH (6.8). Thus water-insoluble drug carriers from natural proteins can be used as matrices for modulated release [65]. Furthermore, Gracia *et al.* used apple citrus pectin to develop biodegradable matrix based microspheres aerogels containing maghemite nanoparticles that were both cylindrical and spherical in shape. The maghemite nanoparticles were preferentially adsorbed on the surface of aerogel matrix microspheres but their magnetic properties were retained even after getting incorporated into the aerogel matrix. Hence the system can be used for targeted delivery [66]. For dermal drug delivery Guenther *et al.* produced silica aerogels system of dithranol to enhance its dermal availability. The researchers concluded that the drug incorporated in a non-crystalline state in the silica aerogel matrix could improve both the release and the penetration properties of dithranol [67].

3.6   Polysaccharide nanocomposites

Nanocomposites comprise of either a single nanomaterial or several nanomaterials embedded in a bulk material. A nanomaterial can be either 'soft' or 'hard". The nanocomposites can be made of a combination of either two 'soft' nanomaterials or two 'hard' nanomaterials or a 'hard' and a 'soft' nanomaterial. Polysaccharides such as chitosan, heparin, cellulose, hyaluronan, alginate, pectin, guar, starch/chitosan, chitosan/heparin, cellulose and chitin are used for targeted, tumor therapy, bone tissue regeneration, monoclonal antibody; form a 'soft' nanomaterial. Various methods to prepare the nanocomposites include dip coating, film casting, electrospinning, physical mixing, ionotropic gelation, colloidal assembly, layer-by-layer assembly, covalent coupling, co-precipitation and *in situ* preparation. These methods are based on columbic interactions, hydrogen-bonding, and electrostatic and ionic interactions, and hydrophobic effects.

Trimethyl chitosan nanoparticles were developed by ionotropic gelation for targeting monoclonal antibodies to the tumor site. Likewise, ionotropic gelation of heparin loaded hyluronic acid-chitosan nanoparticles has been investigated for asthma therapeutics. The co-precipitation method has been reported for development of pectin-coated, iron oxide magnetic nanocomposite for removal of Cu^{2+} [68] and heparin-coated iron oxide nanoparticles for targeted delivery of anticancer drugs [69]. The *in situ* preparation of calcium carbonate nanoparticles, in the presence of cellulose fibers, has been reported. Drop-wise addition of sodium hydroxide to calcium chloride and dimethylcarbonate releases CO_2 and calcium carbonate, while nanoparticles form on the surface of cellulose fibres. Similarly the *in situ* growth of zinc oxide nanoparticles in alginate dispersion has been used to develop antibacterial nanocomposites. Cellulose fiber – silver nanoparticle composites find application for wound healing in clinical setting. The sodium carboxymethyl cellulose- silver nanocomposite films loaded with curcumin showed synergistic antimicrobial activity against *E. coli* attributable to both curcumin and silver, thereby suggesting a promising wound and burn dressing [70].

The polysaccharide nanocomposites can be made using metals/metal oxides, structured carbon, inorganic compounds, biomolecules, polysaccharides and functional polymers. In a report by Abdollahi *et al.* cellulose and montmorillonite nanoparticles were added to alginate dispersion to form inorganic- and organic-reinforced bionanocomposites with characteristic thermal and physicomechanical properties [71]. Mansa *et al.* developed guar gum montmorillonite by using solution intercalation method. They found that the while cellulose enhances surface hydrophobicity, tensile modulus and film tensile strength; montmorillonite increases film hydrophilicity [72]. Graphene-cellulose nanocomposite paper was produced by merging amine-modified nanofibrillated cellulose fibers with reduced graphene oxide sheets. The resultant paper had enhanced electromechanical properties [73]. The nanocomposites of starch/ polyaniline developed by oxidative polymerization of aniline were utilized for removal of dye from an aqueous solution based on the fact that the broken intermolecular hydrogen bonds were freely available for interaction with the dye molecules [74].

3.7   Polysaccharide based nanogels

Nanogel is cross linked network formed by covalent linkage consisting of three dimensional polymer chain with a particle size < 200 nm. Nanogels are categorized into physically and chemically cross linked nanogel; and thermosensitive and pH-sensitive nanogels. The physically cross linked nanogel are formed through non-covalent bonds like hydrogen bonding, hydrophobic-hydrophobic, ionic and hydrophilic-hydrophilic interactions. Chemically cross linked nanogels are developed with different cross linking agents through polymeric chains backbone. The cross linking agents not only alter the swelling behaviour, the pore size and the gel morphology; but they also have a key role in achieving the predetermined release kinetics of the loaded active. The release of drug molecules from a nanogel depends on the nature of interaction between the polymeric molecules that may be hydrophobic interaction, or hydrogen bonding or complexation of drug molecules with the polymeric network. .

Pullulan is a polysaccharide, which is hydrophilic and non-toxic in nature and used to formulate self-assembled nanogel formulations [75]. Kousalova *et al.* discussed modified polysaccharides such as acetylated chondroitin sulphate, cross-linked pullulan, dextran derivatives and acetylated chondroitin sulfate as carrier for the delivery of drugs, proteins, peptides and nucleotides [76]. Ulvan is a sulphated polysaccharide is obtained from seaweed *Ulva*. Its acetylated derivative has been used to enhance the solubility of curcumin by formulating nanogels [77]. It was also found that glycol chitosan, cationic dextran and dextran derivatives such as dextran hydroxyethyl methacrylate based nanogels can be used to deliver therapeutically active agent [78].

3.8   Polysaccharide based quantum dots

Quantum dots are fluorescent, inorganic semi conducting nanocrystals that exhibit exquisite size determined electronic and optical features because of quantum internment. Quantum dots are also described as particles comprising of semiconductor materials like zinc oxide, cadmium sulfide, that range from 2.5 to 100 nm. They are then coated by proteins or some types of ligands for improvement in the solubility. Primarily used for imaging, of lately quantum dots are used to target tumor cells and conjugated to peptides, antibodies as well as folate to enhance targeting. While naked quantum dots are toxic, coating with silk fibroin enhances their biocompatibility. Coated quantum dots are promising platforms for *in vivo* administration. Polysaccharides coated quantum dots have been considered to enhance biocompatibility, targeting ability and stability. For the purpose of imaging, hyaluronic acid quantum dots (HA-QDs) were fabricated by electrostatic reaction between QDs (+vely charged) and HA (-vely charged). The resultant HA-QDs depicted optimal colloidal stability, lower cytotoxicity and size integrity than non-functionalized HA-QDs. In a research report, HA-QDs were fabricated by conjugating the --COOH group of QDs with the $-NH_2$ group of modified HA [79]. Higher modification of HA inflicted negative effect on receptor mediated endocytosis. Thus a modification up to 22 mol% was considered sufficient. Studies showed that HA-QDs effectively responded to the hepatoma and hepatic stellate cells that play a key role in liver pathologies. Therefore, post tail vein injections in rats, resulted in higher build-up of HA-QDs in cirrhotic livers in contrast to normal livers. However, the clearance of HA-QDs was quite slow from the diseased liver. This indicates that HA-QDs act as useful imaging agents for detecting chronic liver diseases. Modifying the QDs by polysaccharides can enhance the stability of QDs for imaging application. Well it is not easy to load enough amounts of active pharmaceutical agents in the quantum dots for drug delivery. To overcome this disadvantage, water soluble QDs loaded chitosan nanospheres were fabricated by ion complexation method. This offers the advantage of high drug loading and has more space as well. When the chitosan QDs were given intratumorally in mice, appearance of strong fluorescence signal at tumor site was observed. The nanospheres showed targeted as well as specific delivery of HER2 sRNA to HER 2 SKBR3 cells of breast cancer [80].

In another study, it was found that encapsulation of Mn- ZnS QDs in chitosan can result in diminutive toxicity, and improvement in the functionalization with folic acid. In a report by Bwatanglang *et al*, the authors integrated the targeting ability of folic acid and imaging capacity of Mn-ZnS QDs into a single unit nanosized delivery system [81]. Folic acid chitosan conjugates (FACS) were fabricated and conjugated with Mn-ZnS QDs to form nanocomposite of FACS-Mn-ZnS QDs. Both, Mn-ZnS and FACS Mn-ZnS nanocomposites showed toxicity against normal breast cells and cancerous breast cells. On increasing the concentration of QDs dots, a minor reduction in the feasibility of cancerous cells served with FACS-Mn-ZnS was recorded in comparison to Mn-ZnS QDs. The reduction is attributable to increased coupling between FA linked QDs and the folate receptor expressing cancer cells. Additionally, it was depicted that FA conjugated QDs emitted strong fluorescence on the binding to the folate receptor expressing cancerous cells.

In another study, cadmium telluride nanocapsules were developed using chondroitin sulfate (anioinic polysaccharide) that were utilized to encapsulate +vely charged oily core and co-loaded by celecoxib and rapamycin [82]. On chondroitin sulfate, a layer of cationic gelatin linked QDs got deposited that retarded the non-specific uptake by normal cells. It was found that; gelatin was hydrolyzed at tumor sites by matrix metalloproteinases. This led to the release of QDs as well as drug nanocapsules into cancerous cells for imaging application. However, when gelatin was replaced with lactoferrin, an ON-OFF effect was observed, where QD fluorescence gets influenced by transfer of energy which later gets stored in tumor cells. Thus hybridization of QDs by polysaccharides results in improved tumor targeting capability.

3.9 Polysaccharide decorated nanoparticles

Nanoparticles are particles of sub-micron size that can easily pass via small capillary due to their ultra small size (<1000 nm) and can avoid rapid clearance by which the residence in the blood stream is prolonged. They can exhibit controlled release behaviour owing to their sensitivity to pH and temperature and biodegradability. Currently, nanoparticles are being used for delivery of drugs, polypeptides, proteins, nucleic acid, genes etc. The materials for nanoparticle preparation include poly(lactic acid), poly (glycolic acid), proteins and peptides and polysaccharide particularly chitosan. Out of all these materials, the most relevant material for preparing nanoparticles for delivery of drug is polysaccharides.

Hydrophilic groups such as amino, carboxyl and hydroxyl present in polysaccharide can form non-covalent bonds with the biological environment to affect bioadhesion. Therefore, when nanoparticles are made by using bioadhesive polysaccharides, it enhances the residence time of the drug loaded nanoparticles. Many reports on nanoparticles of polysaccharides as well as their derivatives for drug delivery can be found in literature. For polysaccharide decorated nanoparticles, four methodologies are generally observed, namely ionic cross linking, polyelectrolyte complexation, covalent cross linking, and self-assemblage of hydrophobically modified polysaccharides. Chitosan is frequently used for the preparation of nanoparticles using glutaraldehyde as cross linker. But glutaraldehyde causes serious toxicity to the cells which limits its uses in the delivery of drug. Therefore, biocompatible covalent cross linking agents can be used. Biocompatible crosslinking agents include carbodiimide, natural di-carboxylic acid, and tri-carboxylic acids (tartaric acid, citric acid and malic acid). The nanoparticles formed by this process were found to be stable. In ionically crosslinked polysaccharide nanoparticles, low molecular weight polycations and polyanions were used as ionic cross-linkers. It has been reported that ionic crosslinking has more advantages than covalent crosslinking as ionic cross linked nanoparticles are mild and prepared through simple procedures. The most extensively used polyanion cross-linker is tripolyphosphate. The first triphosphate crosslinked chitosan was reported by Alonso in 1997. Tripolyphosphate-chitosan nanoparticles have been investigated for delivery of drugs and macromolecules. In polyelectrolyte complexation, the polyelectrolyte polysaccharide forms complexes by intermolecular electrostatic interaction. This can be acquired by regulating the molecular weight of polymers up to a definite range.

Theoretically polyelectrolytes are confined to be biocompatible and water soluble polymers for safety purpose. Ultimately, the only natural polycationic polysaccharide that can meet the criteria is chitosan. Polymers like carboxymethyl cellulose, alginate, dextran sulfate, heparin, etc can be complexed with chitosan to fabricate nanoparticles. In self-assembly of hydrophobically modified polysaccharides, the hydrophobic components are grafted on hydrophilic polymers to get amppiphilic copolymers. When polymeric amphiphiles come in contact with an aqueous environment, micelles or micelle like aggregates are spontaneously formed by intermolecular association within hydrophobic moieties for minimizing interfacial free energy.

In a study by Zhu *et al.* selenium nanoparticles anchored with *Ulva lactuca* polysaccharide were formulated that were significantly effective against acute colitis model in mice. The nanoparticles inhibited NF-κB mediated hyper inflammation [83]. Wang *et al.* showed that metallic nanoparticles were stabilized using various polysaccharide/polysaccharide derivatives for improved antimicrobial effect. They showed that pullulan, pectin, chitosan, starch, chitosan-poly(acrylamide), guar gum, hydroxypropylcellulose, chitosan-carboxymethyl cellulose, xanthan gum, gum ghatti and gum kondagogu are able to deliver active therapeutic agent for better therapeutic effect [84]. Namazi *et al.* discussed that polysaccharide and polysaccharide derivatives can be utilized for the formulation of nanoparticles to deliver therapeutic active agent. Authors found that starch, chitosan, dextran, pullulan, cyclodextrins, cellulose, lactic acid and stearic acid grafted starch, α-cyclodextrin and caprolactone grafted chitosan, PLGA and 1,2- epoxy-3- phenoxypropan grafted dextran, alkyl bromide (octyl, decyl or dodecyl) and acetic anhydride grafted pullulan were utilized to formulate nanoparticles for the delivery of therapeutic agents [85].

3.10   Polysaccharide micelles

Micelles comprises of amphiphilic macromolecules having a hydrophilic shell and hydrophobic core. Spontaneous self-assemblage of amphiphilic molecules above the critical micelle concentration results in micelles formation. In aqueous environment the polar region faces micellar surface and non-polar region forms the micelle core. Therefore, micelles are able to deliver hydrophobic and hydrophobic drugs and the macromolecules. Generally, the particle size of micelles ranges from 5-50 nm depending on the concentration. Basically, micelles that are developed for delivery of drug molecules must be biodegradable, biocompatibile, stable and non-immunogenic. The polysaccharides fulfill the listed requirements and can be modified to exist in charged and neutral states. Additionally, polysaccharides contain variety of reactive groups that include hydroxyl, carboxyl, and amino groups that indicates the possibility of chemical modification in the micelles. The molecular weight of polysaccharide varies from 100 to 1000 Daltons.

A number of functional groups attach to the polysaccharide backbone that facilitate attachment of hydrophobic moiety and initiate self-assembly. The most reliable hydrophobic moieties used for initiation of self-assembly to a micelle are cholesterol, stearic acid, cholic acid, pluronic, polylactide, polylactide-co-glycolide, deoxycholic acid, polycaprolactone. It has been reported by several investigators that cholesterol modified pullulan forms multiple hydrophobic microdomains in place of single hydrophobic case after self-assembly. These multiple microdomains act as physical cross-linkers that enclose the exterior part of hydrophilic shell and ultimately form self-assembled constructs referred to as nanogels. Systems formed by this technique include pullulan based systems, cellulose based systems, dextran based systems, chitosan based systems, heparin based systems, hyaluronan based systems etc. In pullulan based systems, the first reported self aggregated colloidal system is cholesterol bearing pullulan that showed high stability. The cholesterol bearing pullulan was developed by grafting cholesterol groups to various glucose units on pullulan in a random way. The transmission electron microscopy technique confirmed the self aggregation of cholesterol bearing pullulan as uniform spherical structures. The cholesterol bearing pullulan self-aggregates can be loaded with insulin that confer protection to insulin from enzymatic degradation and denaturation. Recently, cholesterol bearing pullulan has been modified and crosslinked with various hydrogels. The crosslinked cholesterol bearing pullulan can be useful in the treatment of bone degeneration; and promotes regeneration of bones. Other pullulan based micelle systems include pullulan acetate, pullulan-g-PLA, PLGA graft pullulan, and pullulan hydrophobic drug conjugates such as pullulan biotin [86, 87].

4. Drug release mechanisms

Gums and mucilages, because of their polysaccharidic nature, possess immense potential to function as hydrophilic matrices from which either water-soluble or water insoluble active pharmaceutical ingredient can be released in a controlled and tailor-made fashion. Swelling in the gums occurs as a result of balance between osmotic pressure, electrostatic pressure and entropy-favored dissolution in the water. This occurs because of porous cross-linked hydrogel structure. Low intermolecular forces with inherent low chain packing in the internal structure, offers little resistance to water permeation and thus exhibit very fast water diffusion. The driving forces for transport of drug molecules through polymer matrix are solvent penetration, swelling, chain disentanglement, relaxation, drug diffusion and erosion/degradation of matrix of which the phenomena of diffusion and erosion contribute significantly. A gel layer or sheath of swollen polymer forms on the surface of the tablet whose thickness increases with time due to alterations in interactions between polymer, drug molecules and water. The gel layer acts as a barrier to subsequent influx of the medium and thus retards release rate of drug. Influx of water/ body fluid leads to swelling of polymer with formation of two distinct phases of polymer-inner glassy phase (crystalline region) and swollen rubbery phase (hydrated/gellified region) characterized by lowering of the glass transition temperature and finally relaxation of the hydrated polymer chains at the surface. In absence of any solute, the primary step of water penetration is governed by diffusion of the solvent and relaxation of the polymeric chains which depend on the water content in the above mentioned regions of the matrix. Relaxation involves rotations, translations and vibrations of the constituent chains. Though the polymer relaxation is the final step it occurs through a series of events. Solvent penetration leads to a change in the configuration in the polymer network which becomes elongated so as to cause swelling and this promotes further water penetration. However, this is opposed by the development of counteracting elastic restrictive forces. Finally, equilibrium is attained when the two forces of elastic restrictive forces and osmotic pressure governing influx of solvent balance each other. The transition from glassy to rubbery state occurs at a critical or threshold polymer concentration with initiation of the processes of chain disentanglement and polymer dissolution. Macromolecular relaxation at the glass-rubbery interface indicates stress generation due to swelling of the polymer, which at the extreme point (severe hydration) initiates the final step of matrix erosion when inter-chain intermolecular forces can no longer withstand any external forces. This emphasizes the significance of the phenomenon of swelling in the process of drug diffusion through the hydrogel matrix. The equilibrium swelling capacity depends on structure of the hydrogel, hydrophilic content, cross-link density and ionic content. Concentration of swelling medium is also a factor of major concern mainly for diffusion of water-soluble drugs since swelling will not occur to maximum capacity in situation of lack of medium availability and hence, diffusion of drug will be less than optimal. Drug initially dissolved/dispersed in the glassy layer diffuses out slowly through the rubbery phase and with time the rate decreases because the thickening of viscous gel layer on the outer surface increases the diffusional path length and offers resistance to the efflux. In case of water-insoluble drugs, the degradation of polymer affects the drug release from the matrix. Drug release from gum- or mucilage-based hydrophilic polymeric matrix can be represented as occurring stepwise through four fronts formed once the solvent penetration induced swelling and erosion of polymer chain have been initiated. Polymer dissolution/erosion is marked by the disappearance of all the four fronts as demarcated in Fig 2.

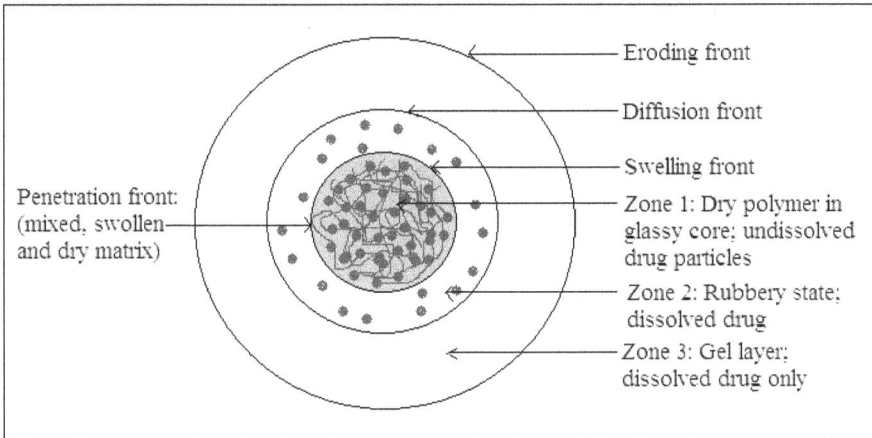

Figure 2. Schematic diagram of polymer dissolution/erosion front.

Eroding front: It demarcates the matrix and the penetrating solvent outside the matrix. The direction of its movement depends on the predominance of erosion and swelling processes. If swelling phenomenon dominates, the boundary between the release environment and the matrix moves outwards. Erosion is governed by the hydrodynamic conditions of the release medium and strength of the network connections. **Swelling front:** Exists as a boundary between inner glassy core and outer swollen rubbery layer, moves inward at a speed determined by the porosity of the matrix. **Diffusion front:** Separates the completely dissolved drug in the outermost part of the rubbery zone and drug yet to be dissolved in the inner layers of the rubbery region and located between swelling and erosion fronts. **Penetration front:** It exists at the interface between glassy core and the hydrated portion of the glassy core [88, 89].

The drug release from gum based matrix systems may be governed by swelling or diffusion or erosion and sometimes a combination of any two phenomena leading to swelling-controlled release systems, diffusion-controlled release systems, chemically controlled systems and environmentally responsive systems. Though swelling and erosion of the matrix have been depicted as two distinct steps, there may be synchronization between these two phenomena occurring simultaneously as has been reported with locust bean gum matrix, thereby maintaining a constant gel layer [90]. Helical structure of the xanthan molecule imparts a degree of order which is disrupted following hydration and thereby enables release of the drug molecule. Initial high degree of swelling due to water penetration followed by low degree of erosion at later stages characterizes the release of drug from xanthan gum based formulations. However, incorporation of locust bean gum in fixed proportions produces synchronization between swelling and erosion of the hydrophilic matrix forming a constant gel layer. The burst phenomenon is controlled by the proportion of xanthan gum in the formulation. Xanthan gum at lower percentage swells up rapidly with less rigid hydrogel structure leading to rapid release of drug. At higher percentage, degree of hydration increases markedly, increasing diffusional path length and hence diminishing the release rate of the entrapped molecule. Minor alteration in the composition of the formulation with respect to xanthan gum can modulate the release pattern substantially. Locust bean gum, if used alone, fails to produce the desired effect of controlled release because of its rapid erosion and loss in structural integrity. The structure of locust bean gum and its erodible nature make it a very versatile candidate for synergistic action in sustained delivery of drug in combination with suitable hydrophilic gums [91].

In a study, it was observed that the swelling of gellan gum was dependent on temperature, the concentration of cation and environmental pH. On exposure to water, gellan gum based microbeads imbibe water, assuming a gel like consistency through which water penetrates or moves inwards and helps in dissolution of the entrapped drug molecules which then diffuse out of the matrix through gel layer. As the drug molecules come out, they create a network of water-filled tortuous channels across the matrix, the extent of which is a direct function of drug load. The initial burst phenomenon observed from the beads at all pH values is attributable to the crystalline drug present on surface of the beads [92].

Gum hakea was found to release the incorporated chlorpheniramine maleate from its matrix by polymeric chain relaxation preceded by hydration [93]. Drug release from the sodium alginate matrix was dependent on the pH of the medium and its swelling may be assumed to depend on the ionic content of the system as well as the surroundings, specially, divalent calcium ion. For such systems, swelling is more entropy favored process at a given amount of elastic forces. The magnitude of osmotic and electrostatic forces increase with the number of ions in the system and thus the matrix thermodynamically behaves like a liquid. Porous hydrated structure is formed due to conversion of the alginate to alginic acid in acidic pH releasing the drug faster whereas neutral pH promotes formation of a viscous gel layer. The profile remains same even if calcium gluconate was added to the matrix, which acts as a pore former although it increases the rigidity of the gel matrix. Since, both locust bean gum and guar gum are non-ionic polysaccharides; drug release from such matrices is independent of pH. It is worthwhile to mention that colon targeting ability of guar gum based systems is in no way related to pH effect but to its specific cleavage and breakdown by the anaerobic colonic bacteria [94]. Drug release from three layer matrix tablet necessitates special consideration. Release retardant polymer present on either sides of the matrix hampers solvent interaction and penetration to the core thus reducing the effective surface area accesible for drug release and burst effect is suitably controlled. With the onset of dissolution, erosion of the gelatinous barrier layer occurs which increases the surface area available for drug release and increases the diffusional path length. Finally, the solvent is able to access the dry unswollen core of the matrix gums and mucilages act as superdisintegrants mainly because of their unique property to swell in aqueous medium due to presence of pores and cross-link joints, reduce intermolecular forces. This activity is manifested at much lower concentration than that required for sustained or modified release effect. Water penetration promotes swelling creating zones of high stress, and as it occurs, an outer isotropic radial pressure is created on the tablet; physically weakening the tablet structural integrity causing its rupture. Thus, the matrix produced by gums keeps its original shape even after swelling. Since, they are added at very low concentrations, characteristic gelling property of the gums does not affect drug release under any circumstances.

5.   Factors affecting drug release

Though the hydrophilic matrix is one of the simplest techniques for regulating rate of drug delivery from a device, mechanisms involved with the ensemble of phenomena ruling release process are quite complex and depend on the interplay of several factors. Profound knowledge of the various factors will help towards realization of the aim of successful controlled drug delivery. The polymer matrix may be assumed to be a percolative network because of the presence of wide network meshes acting as accessible sites for the diffusant and small meshes serving as forbidden sites. Because of this complex network topology and internal high degree of disorder, drug distribution in the matrix strongly influence the release profile and kinetics. Even if the drug is homogeneously distributed throughout the matrix, burst effect may be observed following which drug will be released slowly. Factors affecting the complex release process are physicochemical properties of the solute and the polymer (processes of adsorption/desorption may occur on the matrix surface during drug diffusion), solvent polymer interactions, structural framework and morphology of the matrix, drug-polymer ratio (alteration in the composition may change the tortuosity), ratio between size of diffusant molecule and mesh size, release environment (pH may affect the release rate from ionic polymers), thickness and the strength of the gel layer formed, presence of contaminants like fillers in the formulation The thickness of gel layer depends on the drug load and viscosity of the polymer. Presence, position and movement of the diffusion fronts depend on the drug dose in the matrix [95, 96].

6.   Kinetics of drug release

Mathematical modeling of the release data enables elucidation of mechanisms of solute transport processes through the natural gum based matrix type delivery devices. It also facilitates prediction of release kinetics even before the drug release is realized in practice. Finally, it plays a key role in process optimization of complex drug delivery systems. Since, in polysaccharide based drug delivery systems, drug release is primarily controlled by swelling and relaxation of the hydrophilic polymeric chains preceded by water-uptake, such drug delivery systems may be aptly recounted by two dimensionless numbers: (i) Diffusional Deborah (De) number and, (ii) Swelling Interface number. The magnitude of De ($>>1$ or $<< 1$) indicates whether the swelling process is dominated by polymer relaxation or water penetration and the entire drug diffusion process can be explained by Fick's first law of diffusion. Characteristic diffusion time can be smaller than the polymer relaxation time when the solvent is absorbed before the polymer gets a chance to completely relax [97, 98].

7.   Future prospective

Cellulose, starch and its derivatives, chitin and chitosan have become standard excipients during the last few decades. In future, the polysaccharides based drug delivery systems will occupy stronger position. The marine polysaccharides are emerging as important pharmaceutical excipients of drug delivery systems. Examples of marine polysaccharides are chitosan, chitin and its derivatives chitosan, agar, alginate and collagen. Over the last decades, many innovative formulations have been explored to negotiate drug release for longer periods. Recently, a novel method was developed to allow prolonged release and plasmid DNA expression *in vitro.* Collagen and its derivatives were used in this system that acted as carrier for prolonged delivery of DNA vectors. The chitosan based modified release dosage forms suffer from fast dissolution in stomach. Since chitosan is positively charged at low pH values of stomach it spontaneously reacts with negatively charged polyions to form polyelectrolyte complexes. Researchers need to address this challenge.

As a well documented fact the concentration of the colonic microflora gets altered in the disease conditions. Consequently the degradation profile of natural polymers that are degraded by the colonic bacteria should be studied in its disease state for designing colonic drug delivery systems. The future challenge would be to find a polysaccharide that would from a non-permeable film and at the same time present high degradability. The polysaccharides with low water solubility offer better drug retaining capability, but at the same time suffer from poor degradability. As the colonic enzymes are inducible, it may be possible to induce the select enzyme(s) by achieving colonic delivery of the polysaccharide prior to the drug delivery system.

The commercial sustainability is evidenced by patents utilizing polysaccharide for the delivery of therapeutic agents is tabulated as Table 2.

Table 2. A cross section of patents on polysaccharide based drug delivery.

Patent number	Year	Highlight
US15/59 3,871	2019	The work describes the mechanism of polysaccharide microspheres formation that includes contacting a solvent with a modified cellulose to form a solution; contacting the modified cellulose solution with contacting discontinuous phase of liquid with a continuous phase liquid; at least one bioactive agent to form a discontinuous phase liquid to form an emulsion; and contacting the emulsion, with a third phase liquid to extract the solvent from the emulsion, thereby forming a plurality of modified cellulose microspheres [99].
US16/09 7941	2019	They developed the kits, compositions and methods for modifying for altering polysaccharide fillers and drug delivery systems with the application of an ascorbate [100].
US15/11 0634	2018	They worked on the field of conjugate vaccines of pneumococcal capsular saccharide. Specifically, a multivalent streptococcus pneumoniae immunogenic composition is allocated with different conjugated capsular saccharides from various S. pneumoniae serotypes conjugated two or more different carrier proteins. Polysaccharide is able to deliver diphtheria toxoid for therapeutic effect [101].
US14/72 9408	2018	A tunable vancomycin loaded hydrogel formulation was made. To obtain desired gel strength the ion concentration in the gel material was altered. The hydrogel formulation was layered directly upon burnt skin. The hydrogel formulation also included microparticles and/or nanoparticles such as activated carbon powder that has absorbed additional antibiotic. The particles are used to aid in attaining a timely or sustained release of the antibiotic drug [102].
US14/94 2,435	2018	The invention embodies the excipient combination of rice starch and cellulose or a derivative thereof. The rice starch is pregelatinised or in the form of grains and the cellulose is crystalline or powered. In some quality the composition is in the form of tablets or capsules. Composition preferably does not contain dyes, inks or colourings and are free from sulphites, benzoates and parabens. Carriers and drug delivery systems for the composition are also disclosed [103].

US14/41 1,928	2018	The invention relates to a process for preparing a polysaccharide derivative, comprising the steps of: (a) contacting at least one polysaccharide with at least one polysaccharide swelling agent at a temperature of at most 70.degree. C.; and (b) subsequently, contacting the product of step (a) with at least one aromatic isocyanate; thereby preparing a polysaccharide derivative. The invention deals with modification of polysaccharides at about 70° C. Superior properties isocyanate derivatives of polysaccharide were prepared [104].
UK1717 991.2	2018	The inventors activated *streptococcus pheumoniae* serotype 10A, 22F, or 33F polysaccharides. They also relates to immunogenic conjugates comprising Streptococcus pneumoniae serotype 10A, 22F or 33F polysaccharides covalently linked to a carrier protein, processes for their preparation and immunogenic compositions and vaccines comprising them. For better therapeutic effect covalently linked polysaccharide [105].
US15/18 31077	2017	The invention embodies a multiple pulsed dose drug delivery system for pharmaceutically active amphetamine salts, comprising a pharmaceutically active amphetamine salt covered with an immediate-release coating and a pharmaceutically active amphetamine salt covered with an enteric coating wherein the immediate release coating and the enteric coating provide for multiple pulsed dose delivery of the pharmaceutically active amphetamine salt. The product can be composed of either one or a number of beads in a dosage form, including capsule, tablet, or sachet method for administering the beads [106].
US15/48 0,021	2016	A biomaterial comprising injectable bioresorbable polysaccharides composition where as the polysaccharide may be succino- chitosan glutamate was disclosed. The particles comprising or consisting essentially of chitin and /or chitosan, which may be free of any additional formulation modifying agents, and a process for manufacturing the same. Medical device are relates to a biomaterial and to a method of repair or treatment including the filling of cavities in the human face or body [107].
US15/09 1,142	2016	A method and composition for targeted delivery of substances to the lower GI tract comprises a base or scaffold carrier to which is fixed or covalently attached to a drug or prodrug. When the compound taken orally travels through the GI tract of a patient to the lower GI where bacterial azo reductase enzymes cleave the bonds, releasing the drug or pro drug from the base or scaffold carrier permitting the delivery of the drug to the vicinity of a target cell type in the GI tract. The base or scaffold, which remains as a by- product passes out of the GI tract in the feces [108].
US14/27 1,251	2015	Novel water stable pharmaceutical compositions, their liquid form oral pharmaceutical compositions and kits thereof, rehydration beverages containing these water stable pharmaceutical compositions methods of manufacture and methods of use thereof are disclosed. The novel aqueous delivery systems are useful, as alternative pharmaceutical dosing agents to tablets, capsules and other forms of delivering medication to a mammalian host in need thereof [109].

US12/68 1,676	2015	The invention embodies a method of tissue engineering for a porous scaffold. It discloses the method and its use for tissue engineering, cell culture and cell delivery. Consisting of steps a) preparing an alkaline aqueous solution comprising an amount of at least one polysaccharide and one cross-linking agent b) freezing the aqueous solution of step a) c) sublimating the frozen solution of step b). Characterized in that step b) is performed before the cross-linking of the polysaccharide occurs in the solution of step a) [110].
US13/49 8,621.	2015	The invention discloses particulate polysaccharides derivative production by dry- grinding a moist polysaccharide derivative. [111].
US13/23 1,150	2014	The invention discloses pharmaceutical compositions for target delivery of a drug to cells expressing higher levels of P-glycoprotein compared to other cells in a mammal, for the treatment of various diseases or conditions, such as cancer and neurological conditions [112].
US15/18 3/077	2013	The aldehyde-functionalized polysaccharides may be reacted with various amine-containing polymers to form hydrogel tissue adhesives and sealants that may be useful for medical applications such as wound closure, intestinal anastomosis and vascular anastomosis, issue repair, preventing leakage of fluids such as blood, bile, gastrointestinal fluid and cerebrospinal fluid, ophthalmic procedures, drug delivery, and preventing post-surgical adhesions [113].
US12/80 9,629	2011	Describes a liquid controlled– release drug delivery compositions which gel upon injection into the body to form, in situ, controlled –release drug implants. The compositions of gel- forming polymer feature that is insoluble in water, a polyethylene glycol solvent in which the polymer is dissolved, and the drug substance to be delivered [114].
US10/74 0,436	2010	The invention discloses the process of isolation of highly-branched glucose homopolymers from various organisms including, but not limited to, microorganisms such as bacteria and yeasts. Also provided are methods for chemical conjugation of the polysaccharide nanoparticles with various agents. Also provides examples of use of the polysaccharide nanoparticles and their derivatives as drug delivery systems and fluorescent diagnostics [115].
US11/99 2,272	2009	The invention discloses a method of delivering a therapeutic agent by providing a cross- linked polymer encapsulating the therapeutic agent to a site in a patient. The cross-linked polymer degradation rate is correlated with a local concentration of an indicator, and the therapeutic agent is released as the cross-linked polymer degrades [116].
PCT/I L200 4/000 123	2006	It discloses a delivery system incorporating pharmaceutical products for delivery to an eye in the treatment of primary open –angle glaucoma. The formulation comprises (a) a biocompatible erodible material incorporating a therapeutically-effective amount of a prostaglandin analogue, and (b) a reservoir containing a therapeutically- effective amount of a beta-blocker, whereby, when the delivery system is placed in the eye the prostaglandin analogue is delivered gradually as the erodible material is eroded, and the beta-blocker is delivered rapidly when at least a predetermined portion of the erodible material has been eroded [117].

References

[1] A.M. Patten, D.G. Vassao, M.P. Wolcott, L.B. Davin, N.G. Lewis, Trees: a remarkable biochemical bounty. In Comprehensive Natural Products II: Chemistry and Biology, Elsevier Ltd. 2010, pp. 1173-1296. https://doi.org/10.1016/B978-008045382-8.00083-6

[2] V.D. Prajapati, G.K. Jani, N.G. Moradiya, N.P. Randeria, Pharmaceutical applications of various natural gums, mucilages and their modified forms, Carbohydr Polym. 92(2013) 1685-99. https://doi.org/10.1016/j.carbpol.2012.11.021

[3] J. Joseph, S.N. Kanchalochana, G. Rajalakshmi, V. Hari, R.D. Durai, Tamarind seed polysaccharide: A promising natural excipient for pharmaceuticals, Int J Green Pharm. 6 (2012) 270-278. https://doi.org/10.4103/0973-8258.108205

[4] A.L. Harvey, Natural products in drug discovery, Drug Discov Today. 13(2008) 894-901. https://doi.org/10.1016/j.drudis.2008.07.004

[5] V. Rana, P. Rai, A.K. Tiwari, R.S. Singh, J.F. Kenedy, C.J. Knill, Modified gums: approaches and application in drug delivery, Carbohydr Polym. 83 (2011) 1031-1047. https://doi.org/10.1016/j.carbpol.2010.09.010

[6] R. Khathuriya, T. Nayyar, S. Sabharwal, U.K. Jain, R. Taneja, Recent aproaches and pharmaceutical applications of Natural Polysaccharides: A Review, Int. J. Pharm. Sci. Res. 6 (2015) 4904-4919.

[7] J. Joseph, S.N. Kanchalochana, G. Rajalakshmi, V. Hari, R.D. Durai, Tamarind seed polysaccharide: A promising natural excipient for pharmaceuticals, Int. J. Green Pharm. 6(4) (2012) 270-278. https://doi.org/10.4103/0973-8258.108205

[8] M.K. Chourasia, S.K. Jain, Polysaccharides for colon targeted drug delivery, Drug Deliv. 11(2) (2004) 129-48. https://doi.org/10.1080/10717540490280778

[9] E. Alpizar-Reyes, H. Carrillo-Navas, R. Romero-Romero, V. Varela-Guerrero, J. Alvarez-Ramírez, C. Pérez-Alonso, Thermodynamic sorption properties and glass transition temperature of tamarind seed mucilage (Tamarindusindica L.), Food and Bioprod. Process. 101 (2017) 166-176. https://doi.org/10.1016/j.fbp.2016.11.006

[10] J.D. Smart, Buccal drug delivery, Expert Opin Drug Deliv. 2(3) (2005) 507-17. https://doi.org/10.1517/17425247.2.3.507

[11] O. Felt, P. Buri, R. Gurny, Chitosan: a unique polysaccharide for drug delivery, Drug Dev. Ind. Pharm. 24(11) (1998) 979-993. https://doi.org/10.3109/03639049809089942

[12]F. Acarturk, Mucoadhesive vaginal drug delivery systems, Recent Pat. Drug Deliv. Formul. 3(3) (2009) 193-205. https://doi.org/10.2174/187221109789105658

[13]A.K. Nayak, D. Pal, Tamarind seed polysaccharide: An emerging excipient for pharmaceutical use, Indian J. Pharm. Educ. Res. 51 (2017) S136-46. https://doi.org/10.5530/ijper.51.2s.60

[14]M.L. Bruschi, O. de Freitas, Oral bioadhesive drug delivery systems, Drug Dev. Ind. Pharm. 31(3) (2005) 293-310. https://doi.org/10.1081/DDC-52073

[15]D.H. Patel, M.P. Patel, M.M. Patel, Formulation and evaluation of drug-free ophthalmic films prepared by using various synthetic polymers, J. Young Pharm. 1(2) (2009) 116-120. https://doi.org/10.4103/0975-1483.55742

[16]Y. Ting-Ting, C. Yuan-Zheng, Q. Meng, W. Yong-Hong, Yu. Hong-Li, W. An-Lin, Z. Wei-Fen. Thermosensitive chitosan hydrogels containing polymeric microspheres for vaginal drug delivery, BioMed Res. Int. (2017) 12. https://doi.org/10.1155/2017/3564060

[17]C. Valenta, The use of mucoadhesive polymers in vaginal delivery. Adv. Drug Deliv. Rev. 57(11) (2005) 1692-1712. https://doi.org/10.1016/j.addr.2005.07.004

[18]M. Sharma, N. Sharma, A. Sharma. Rizatriptan benzoate loaded natural polysaccharide based microspheres for nasal drug delivery system. Int. J. App. Pharm. 10(5) (2018) 261-269. https://doi.org/10.22159/ijap.2018v10i5.27877

[19]M.N. Jones, Carbohydrate-mediated liposomal targeting and drug delivery, Adv. Drug. Deliv. Rev. 13 (1994) 215-249. https://doi.org/10.1016/0169-409X(94)90013-2

[20]J. Sunamoto, Application of Polysaccharide-coated Liposomes in Chemotherapy and Immunotherapy. In Medical application of liposomes Karger Publishers. 1986, pp. 121-129. https://doi.org/10.1159/000413501

[21]J. Sunamoto, K. Iwamoto, H. Kondo, Liposomal membranes. VII. Fusion and aggregation of egg lecithin liposomes as promoted by polysaccharides, Biochem Biophysic Res Comm. 94 (1980) 1367-1373. https://doi.org/10.1016/0006-291X(80)90570-7

[22]J. Sunamoto, K. Iwamoto,Protein-coated and polysaccharide-coated liposomes as drug carriers, Crit Rev Ther Drug Carrier Syst. 2 (1986)117-136.

[23]J. Moellerfeld, W. Prass, H. Ringsdorf, H. Hamazaki, J. Sunamoto, Improved stability of black lipid membranes by coating with polysaccharide derivatives bearing

hydrophobic anchor groups, Biochim Biophys Acta Biomembr. 857 (1986) 265-270. https://doi.org/10.1016/0005-2736(86)90355-X

[24] T. Sato, J. Sunamoto, Recent aspects in the use of liposomes in biotechnology and medicine, Prog Lipid Res. 31 (1992) 345-372. https://doi.org/10.1016/0163-7827(92)90001-Y

[25] M.G. Elferink, J.G. Wit, G. In't Veld, A. Reichert, A.J. Driessen, H. Ringsdorf, W.N. Konings, The stability and functional properties of proteoliposomes mixed with dextran derivatives bearing hydrophobic anchor groups, Biochim Biophys Acta Biomembr. 1106 (1992) 23-30. https://doi.org/10.1016/0005-2736(92)90217-A

[26] P. Deol, G.K. Khuller, Lung specific stealth liposomes: stability, biodistribution and toxicity of liposomal antitubercular drugs in mice, Biochim Biophys Acta Gen Subj. 1334 (1997) 161-172. https://doi.org/10.1016/S0304-4165(96)00088-8

[27] H. Takeuchi, H. Yamamoto, T. Niwa, T. Hino, Y. Kawashima, Mucoadhesion of polymer-coated liposomes to rat intestine *in vitro*,Chem. Pharm. Bull. 42 (1994) 1954-1956. https://doi.org/10.1248/cpb.42.1954

[28] H. Takeuchi, H. Yamamoto, T. Niwa, T. Hino, Y. Kawashima, Enteral absorption of insulin in rats from mucoadhesive chitosan-coated liposomes, Pharm Res. 13 (1996) 896-901. https://doi.org/10.1023/A:1016009313548

[29] G. Liang, Z. Jia-Bi, X. Fei, N. Bin, Preparation, characterization and pharmacokinetics of N-palmitoyl chitosan anchored docetaxel liposomes, J Pharm Pharmacol. 59 (2007) 661-667. https://doi.org/10.1211/jpp.59.5.0006

[30] Y. Wang, S. Tu, R. Li, X. Yang, L. Liu, Q. Zhang, Cholesterolsuccinyl chitosan anchored liposomes: preparation, characterization, physical stability, and drug release behaviour, Nanomed Nanotechnol Biol Med. 6 (2010) 471-477. https://doi.org/10.1016/j.nano.2009.09.005

[31] H.W. Tan, M. Misran, Polysaccharide-anchored fatty acid liposome, Int J Pharm. 441 (2013) 414-423. https://doi.org/10.1016/j.ijpharm.2012.11.013

[32] G.B. Jiang, D. Quan, K. Liao, H. Wang, Preparation of polymeric micelles based on chitosan bearing a small amount of highly hydrophobic groups, Carbohydr Polym. 66 (2006) 514-520. https://doi.org/10.1016/j.carbpol.2006.04.008

[33] M. Othman, K. Bouchemal, P. Couvreur, R. Gref, Microcalorimetric investigation on the formation of supramolecular nanoassemblies of associative polymers loaded with

gadolinium chelate derivatives, Int J Pharm. 379 (2009) 218-225.
https://doi.org/10.1016/j.ijpharm.2009.05.061

[34] N. Yerushalmi, R. Margalit, Hyaluronic acid-modified bioadhesive liposomes as local drug depots: effects of cellular and fluid dynamics on liposome retention at target sites, Arch. Biochem. Biophys. 349 (1998) 21-26.
https://doi.org/10.1006/abbi.1997.0356

[35] I. Bataille, J. Huguet, G. Muller, G. Mocanu, A. Carpov, Associative behaviour of hydrophobically modified carboxymethyl pullulan derivatives, Int J Biol Macromol. 20 (1997) 179-191. https://doi.org/10.1016/S0141-8130(97)01158-6

[36] K. Glinel, J. Huguet, G. Muller, Comparison of the associating behaviour between neutral and anionic alkyl perfluorinated pullulan derivatives, Polymer. 40 (1999) 7071-7081. https://doi.org/10.1016/S0032-3861(99)00085-3

[37] I. Lalush, H. Bar, I. Zakaria, S. Eichler, E. Shimoni, Utilization of amylose–lipid complexes as molecular nanocapsules for conjugated linoleic acid, Biomacromol. 6(2005) 121-130. https://doi.org/10.1021/bm049644f

[38] H. Ayame, N. Morimoto, K. Akiyoshi, Self-assembled cationic nanogels for intracellular protein delivery, Bioconjugate Chem. 19 (2008) 882-890.
https://doi.org/10.1021/bc700422s

[39] Y. Cao, Y. Gu, H. Ma, J. Bai, L. Liu, P. Zhao, H. He, Self-assembled nanoparticle drug delivery systems from galactosylated polysaccharide–doxorubicin conjugate loaded doxorubicin, Int J BiolMacromol. 46 (2010) 245-249.
https://doi.org/10.1016/j.ijbiomac.2009.11.008

[40] S. Kwon, J.H. Park, H. Chung, I.C. Kwon, S.Y. Jeong, I.S. Kim, Physicochemical characteristics of self-assembled nanoparticles based on glycol chitosan bearing 5β-cholanic acid, Langmuir. 19 (2003) 10188-10193. https://doi.org/10.1021/la0350608

[41] J. Bodillard, G. Pattappa, P. Pilet, P. Weiss, G. Rethore, Functionalisation of polysaccharides for the purposes of electrospinning: a case study using HPMC and Si-HPMC, Gels. 1 (2015) 44-57. https://doi.org/10.3390/gels1010044

[42] K.Y. Lee, L. Jeong, Y.O. Kang, S.J. Lee, W.H. Park, Electrospinning of polysaccharides for regenerative medicine, Adv. Drug Deliv. Rev. 61 (2009) 1020-1032. https://doi.org/10.1016/j.addr.2009.07.006

[43] W. Zhang, D. Hua, S. Ma, Z. Chen, Y.Wang, F. Zhang,F. Len, X. Pu, Preliminary study for vascular tissue engineering by Electrospinning angelica polysaccharide

(ASP)/PLA microfibrous scaffolds, Int. J. Polym. Mater. Po. 63 (2014) 672-679. https://doi.org/10.1080/00914037.2013.854241

[44] A.V. Il'ina, V.P. Varlamov, Chitosan-based polyelectrolyte complexes: A review, Appl Biochem Microbiol. 41 (2005) 5-11. https://doi.org/10.1007/s10438-005-0002-z

[45] Y. Luo, Q. Wang, Recent development of chitosan-based polyelectrolyte complexes with natural polysaccharides for drug delivery, Int J BiolMacromol. 64 (2014) 353-367. https://doi.org/10.1016/j.ijbiomac.2013.12.017

[46] S.K. Motwani, S. Chopra, S. Talegaonkar, K. Kohli, F.J. Ahmad, R.K. Khar, Chitosan–sodium alginate nanoparticles as submicroscopic reservoirs for ocular delivery: Formulation, optimisation and in vitro characterisation, Eur J Pharm Biopharm. 68 (2008) 513-525. https://doi.org/10.1016/j.ejpb.2007.09.009

[47] P. Erbacher, S. Zou, T. Bettinger, A.M. Steffan, J.S. Remy, Chitosan-based vector/DNA complexes for gene delivery: Biophysical characteristics and transfection ability, Pharm Res. 15 (1998) 1332-1339. https://doi.org/10.1023/A:1011981000671

[48] Z. Teng, Y. Luo, Q. Wang, Carboxymethyl chitosan–soy protein complex nanoparticles for the encapsulation and controlled release of vitamin D3, Food Chem. 41 (2013) 524-532. https://doi.org/10.1016/j.foodchem.2013.03.043

[49] H. Wang, M. Roman, Formation and properties of chitosan– cellulose nanocrystal polyelectrolyte– macroion complexes for drug delivery applications, Biomacromol. 12 (2011) 1585-1593. https://doi.org/10.1021/bm101584c

[50] J. Du, R. Sun, S. Zhang, T. Govender, L.F. Zhang, C.D. Xiong, Y.X. Peng, Novel polyelectrolyte carboxymethyl konjac glucomannan–chitosan nanoparticles for drug delivery, Macromol. Rapid Commun. 54 (2004) 954-958. https://doi.org/10.1002/marc.200300314

[51] G. Pasparakis, N. Bouropoulos, Swelling studies and in vitro release of verapamil from calcium alginate and calcium alginate–chitosan beads, Int J Pharm. 323 (2006) 34-42. https://doi.org/10.1016/j.ijpharm.2006.05.054

[52] J.H. Hamman, Chitosan based polyelectrolyte complexes as potential carrier materials in drug delivery systems, Marine drugs. 8 (2010) 1305-1322. https://doi.org/10.3390/md8041305

[53] H.V. Saether, H.K. Holme, G. Maurstad, O. Smidsrod, B.T. Stokke, Polyelectrolyte complex formation using alginate and chitosan, Carbohydr Polym. 74 (2008) 813-821. https://doi.org/10.1016/j.carbpol.2008.04.048

[54] F. Bigucci, B. Luppi, T. Cerchiara, M. Sorrenti, G. Bettinetti, L. Rodriguez, V. Zecchi, Chitosan/pectin polyelectrolyte complexes: selection of suitable preparative conditions for colon-specific delivery of vancomycin, Eur J Pharm Sci. 35 (2008) 435-441. https://doi.org/10.1016/j.ejps.2008.09.004

[55] C. Tapia, Z. Escobar, E. Costa, J. Sapag-Hagar, F. Valenzuela, C. Basualto, M.N. Gai, M. Yazdani-Pedram, Comparative studies on polyelectrolyte complexes and mixtures of chitosan–alginate and chitosan–carrageenan as prolonged diltiazem clorhydrate release systems, Eur J Pharm Biopharm. 57 (2004) 65-75. https://doi.org/10.1016/S0939-6411(03)00153-X

[56] X.Z. Shu, K.J. Zhu, A novel approach to prepare tripolyphosphate/chitosan complex beads for controlled release drug delivery, Int J Pharm. 201 (2000) 51-58. https://doi.org/10.1016/S0378-5173(00)00403-8

[57] J.S. Maciel, D.A. Silva, H.C. Paula, R.C. De Paula, Chitosan/carboxymethyl cashew gum polyelectrolyte complex: synthesis and thermal stability, Eur Polym J. 41 (2005) 2726-2733. https://doi.org/10.1016/j.eurpolymj.2005.05.009

[58] A. Masotti, F. Bordi, G. Ortaggi, F. Marino, C. Palocci, A novel method to obtain chitosan/DNA nanospheres and a study of their release properties, Nanotechnology. 19 (2008) 055302. https://doi.org/10.1088/0957-4484/19/05/055302

[59] J. Stergar, U. Maver, Review of aerogel-based materials in biomedical applications, J Sol-Gel Sci Technol. 77 (2016) 738-752. https://doi.org/10.1007/s10971-016-3968-5

[60] F. De Cicco, P. Russo, E. Reverchon, C.A. Garcia-Gonzalez, R.P. Aquino, P. Del Gaudio,Prilling and supercritical drying: A successful duo to produce core-shell polysaccharide aerogel beads for wound healing, Carbohydr Polym. 147 (2016) 482-489. https://doi.org/10.1016/j.carbpol.2016.04.031

[61] F. Quignard, R. Valentin, F. Di Renzo, Aerogel materials from marine polysaccharides, New J. Chem. 32 (2008) 1300-1310. https://doi.org/10.1039/b808218a

[62] G. Tkalec, Z. Knez, Z. Novak, Formation of polysaccharide aerogels in ethanol, RSC Adv. 5 (2015) 77362-77371. https://doi.org/10.1039/C5RA14140K

[63] M. Alnaief, M.A. Alzaitoun, C.A. Garcia-Gonzalez, I. Smirnova, Preparation of biodegradable nanoporous microspherical aerogel based on alginate, Carbohydr Polym. 84 (2011) 1011-1018. https://doi.org/10.1016/j.carbpol.2010.12.060

[64] M. Alnaief, I. Smirnova, In situ production of spherical aerogel microparticles, J. Supercrit. Fluid. 55 (2011) 1118-1123. https://doi.org/10.1016/j.supflu.2010.10.006

[65] M. Betz, C.A. Garcia-Gonzalez, R.P. Subrahmanyam, I. Smirnova, U. Kulozik, Preparation of novel whey protein-based aerogels as drug carriers for life science applications, J. Supercrit. Fluid. 72 (2012) 111-119. https://doi.org/10.1016/j.supflu.2012.08.019

[66] C.A. Garcia-Gonzalez, E. Carenza, M. Zeng, I. Smirnova, A. Roig, Design of biocompatible magnetic pectin aerogel monoliths and microspheres, RSC Adv. 2 (2012) 9816-9823. https://doi.org/10.1039/c2ra21500d

[67] U. Guenther, I. Smirnova, R.H. Neubert, Hydrophilic silica aerogels as dermal drug delivery systems–Dithranol as a model drug, Eur. J. Pharm. Biopharm. 69 (2008) 935-942. https://doi.org/10.1016/j.ejpb.2008.02.003

[68] J.L. Gong, X.Y. Wang, G.M. Zeng, L. Chen, J.H. Deng, X.R. Zhang, Q.Y. Niu, Copper (II) removal by pectin–iron oxide magnetic nanocomposite adsorbent, Chem. Eng. J. 185 (2012) 100-107. https://doi.org/10.1016/j.cej.2012.01.050

[69] A. Javid, S. Ahmadian, A.A. Saboury, S.M. Kalantar, S. Rezaei-Zarchi, Novel biodegradable heparin-coated nanocomposite system for targeted drug delivery, RSC Adv. 4 (2014) 13719-13728. https://doi.org/10.1039/C3RA43967D

[70] R.J. Pinto, P.A. Marques, C.P. Neto, T. Trindade, S. Daina, P. Sadocco, Antibacterial activity of nanocomposites of silver and bacterial or vegetable cellulosic fibers, Acta Biomaterialia. 5 (2009) 2279-2289. https://doi.org/10.1016/j.actbio.2009.02.003

[71] M. Abdollahi, M. Alboofetileh, M. Rezaei, R. Behrooz, Comparing physico-mechanical and thermal properties of alginate nanocomposite films reinforced with organic and/or inorganic nanofillers, Food Hydrocoll. 32(2013) 416-424. https://doi.org/10.1016/j.foodhyd.2013.02.006

[72] R. Mansa, C. Detellier, Preparation and characterization of guar-montmorillonite nanocomposites, Materials. 6 (2013) 5199-5216. https://doi.org/10.3390/ma6115199

[73] N.D. Luong, N. Pahimanolis, U. Hippi, J.T. Korhonen, J. Ruokolainen, L.S. Johansson, J.D. Nam, J. Seppala, Graphene/cellulose nanocomposite paper with high electrical and mechanical performances, J. Mater. Chem. 21 (2011) 13991-13998. https://doi.org/10.1039/c1jm12134k

[74] V. Janaki, K. Vijayaraghavan, B.T. Oh, K.J. Lee, K. Muthuchelian, A.K. Ramasamy, S. Kamala-Kannan, Starch/polyaniline nanocomposite for enhanced removal of reactive dyes from synthetic effluent, Carbohydr Polym. 90 (2012) 1437-1444. https://doi.org/10.1016/j.carbpol.2012.07.012

[75] T. Zhanga, R. Yanga, S. Yanga, J. Guana, D. Zhanga, Y. Mab, H. Liua, Research progress of self-assembled nanogel and hybrid hydrogel systems based on pullulan derivatives, Drug Deliv. 25 (2018) 278–292. https://doi.org/10.1080/10717544.2018.1425776

[76] J. Kousalova, T. Etrych. Polymeric nanogels as drug delivery systems, Physiol. Res. 67 (2018) S305-S317. https://doi.org/10.33549/physiolres.933979

[77] T.H. Bang, T.T. Thanh Van, L.X. Hung, B.M. Ly, N.D. Nhut, T.T. Thu Thuy, B.T. Huy, Nanogels of acetylated ulvan enhance the solubility of hydrophobic drug curcumin, Bull. Mater. Sci. 42 (2019) 1-7. https://doi.org/10.1007/s12034-018-1682-3

[78] G. Soni, K.S. Yadav, Nanogels as potential nanomedicine carrier for treatment of cancer: a mini review of the state of the art, Saudi Pharm J. 24 (2016) 133–139. https://doi.org/10.1016/j.jsps.2014.04.001

[79] S.H. Bhang, N. Won, T.J. Lee, H. Jin, J. Nam, J. Park, H. Chung, H.S. Park, Y.E. Sung, S.K. Hahn, B.S. Kim, S.Kim, Hyaluronic acid-quantum dot conjugates for in vivo lymphatic vessel imaging, ACS Nano. 3 (2009) 1389-1398. https://doi.org/10.1021/nn900138d

[80] W.B. Tan, S. Jiang, Y. Zhang, Quantum-dot based nanoparticles for targeted silencing of her2/neu gene via RNA interference, Biomaterials. 28 (2007) 1565-1571. https://doi.org/10.1016/j.biomaterials.2006.11.018

[81] I.B. Bwatanglang, F. Mohammad, N.A. Yusof, Folic acid targeted Mn:ZnS quantum dots for theranostic applications of cancer cell imaging and therapy, Int. J. Nanomed. 11 (2016) 413-428. https://doi.org/10.2147/IJN.S90198

[82] A.S. Abdelhamid, M.W. Helmy, S.M. Ebrahim, Layer-by-layer gelatin/chondroitin quantum dots-based nanotheranostics: combined rapamycin/celecoxib delivery and cancer imaging, Nanomedicine (Lond.).13 (2018) 1707–1730. https://doi.org/10.2217/nnm-2018-0028

[83] C. Zhu, S. Zhang, C. Song, Y. Zhang, Q. Ling, P.R. Hoffmann, J. Li, T. Chen, W. Zheng, Z. Huang, Selenium nanoparticles decorated with Ulva lactuca polysaccharide potentially attenuate colitis by inhibiting NF-κB mediated hyper

inflammation, J Nanobiotechnol. 15 (2017) 1-15. https://doi.org/10.1186/s12951-016-0241-6

[84] C. Wang, X. Gao, Z. Chen, Y. Chen, H. Chen, Preparation, characterization and application of polysaccharide-based metallic nanoparticles: A review, Polymers. 9 (2017) 1-34.https://doi.org/10.3390/polym9010001

[85] http://cdn.intechopen.com/pdfs/36882/InTechNanoparticles_based_on_modified_polysaccharides.pdf (accessed on 10 May 2019).

[86] K. Akiyoshi, S. Yamaguchi, J. Sunamoto, Self-aggregates of hydrophobic polysaccharide derivatives, Chem. Lett. 20 (1991) 1263–1266. https://doi.org/10.1246/cl.1991.1263

[87] K. Akiyoshi, S. Kobayashi, S. Shichibe, D. Mix, M. Baudys, S.W. Kim, J. Sunamoto, Self-assembled hydrogel nanoparticle of cholesterol-bearing pullulan as a carrier of protein drugs: Complexation and stabilization of insulin, J. Contr. Release. 54 (1998) 313–320.

[88] H. Omidian, K. Park, Swelling agents and devices in oral drug delivery, J. Drug. Deliv. Sci. Tech. 18 (2008) 83-93. https://doi.org/10.1016/S1773-2247(08)50016-5

[89] M. Grassi, G. Grassi, Mathematical modelling and controlled drug delivery: matrix systems. Current. Drug. Deliv. 2 (2005) 97-116. https://doi.org/10.2174/1567201052772906

[90] J. M. Sankalia, G.M. Sankalia, R.C. Mashru, Drug release and swelling kinetics of directly compressed glipizide sustained-release, J Contr Rel. 129 (2008) 49–58. https://doi.org/10.1016/j.jconrel.2008.03.016

[91] K.S. Rajesh, M.P. Venkataraju, D.V. Gowda, Hydrophilic natural gums in formulation of oral-controlled release matrix tablets of propranolol hydrochloride, J Pharm Sci. 22 (2009) 211-219.

[92] P. Matricardi, C. Cencetti, R. Ria, F. Alhaique, T. Coviello, Preparation and characterization of novel gellan gum hydrogels suitable for modified drug release, Molecules. 14 (2009) 3376-3391. https://doi.org/10.3390/molecules14093376

[93] H.H. Alur, S.I. Pather, A.K. Mitra, T.P. Johnston, Evaluation of the gum from hakea gibbosa as a sustained-release and mucoadhesive component in buccal tablets, Pharm Develop Technol. 4 (1999) 347-358. https://doi.org/10.1081/PDT-100101370

[94] V. Pillay, C.M. Dangor, T. Govender, K.R. Moopanar, N. Hurbans,Drug release modulation from cross-linked calcium alginate microdiscs, 1: evaluation of the concentration dependency of sodium alginate on drug entrapment capacity,

morphology, and dissolution rate, Drug Deliv. 5 (1998) 25-34.
https://doi.org/10.3109/10717549809052024

[95] W. Leobandung, H. Ichikawa, Hydrogels in pharmaceutical formulations, Eur. J. Pharm. Biopharm. 50 (2000) 27-46. https://doi.org/10.1016/S0939-6411(00)00090-4

[96] D.A. Edwards, Non-Fickiandiffusion in thin polymer films, J Polym Sci. 34 (1996) 981-997. https://doi.org/10.1002/(SICI)1099-0488(19960415)34:5<981::AID-POLB16>3.0.CO;2-7

[97] N.A. Peppas, L. Branson, Water diffusion and sorption in amorphous macromolecular systems and foods, J Food Engineer. 22 (1994) 189-210. https://doi.org/10.1016/0260-8774(94)90030-2

[98] J.J. Sahlin, A simple equation for description of solute release. III. Coupling of diffusion and relaxation, Int J Pharm. 57 (1989) 169-172. https://doi.org/10.1016/0378-5173(89)90306-2

[99] P. Blaskovich, R. Ohri, L. Pham, Oxidized cellulose microspheres. US15/593,871. 2019.

[100]   R. Burtt. Kits and methods of using ascorbates to modify polysaccharide fillers and delivery systems. US16/097,941. 2019.

[101]   J. Gu, , Kainthan, R. Kumar, K. Jin-Hwan, A.K. Prasad, Y. Yu-Ying. Streptococcus pneumoniae capsular polysaccharides and conjugates thereof, US15/110634. 2018.

[102]   R. Biemans, L. Marie- Josephe, N. Garcon, P.V. Hermand, P.J. Jan, M.M.P. Van. Pneumoccal polysaccharide conjugate vaccine. US14/729,408. 2018.

[103]   A. Shukla, S. Shukla. Tunable anti-microbial loaded hydrogels, US14/942,435. 2018.

[104]   P. Christopher, H. Servaas, V. Tugba, D. Steve. Process for preparing derivatized polysaccharides. US14/411,928. 2018.

[105]   E. Eren. Hypoallergic drug delivery system. UK1717991.2. 2018.

[106]   A. Forge, D. Vonwill, V.P. Shastri. Method for purifying polysaccharides and pharmaceutical compositions and medical devices containing the same. US15/183/077. 2017.

[107]   A. Shajaee, S. Read, R.A. Couch, P. Hodgkins. Controlled dose drug delivery system. US15/480,021. 2017.

[108] E. Laugier, F. Gouchet. Grandmontagne B; Biomaterial, injectable implant comprising it, its method of preparation and its uses. US15/091,142. 2016.

[109] Mash, E.A. Jr, P.R. Kiela, K. Grishen. Method and compositions for targeted drug delivery to the lower GI tract. US14/271,251. 2016.

[110] C.L. Visage, D. Letourneur, F. Chaubet, A. Autissier. Method for preparing porous scaffold for tissue engineering. US 12/681,676. 2015.

[111] P.E. Pierini, Y.G. Georlach- Doht, J. Hermanns. Process for dry grinding a polysaccharide dervative. US13/498,621. 2015.

[112] K. A. Reed. Aqueous drug delivery system. US13/231,150. 2015.

[113] P. David, C. Lindsey. Drug delivery compositions and methods targeting P-glycoprotein. US15/183,077. 2014.

[114] D. Su, P. Ashton, J. Chen. In situ gelling Drug delivery system. US12/870,616. 2011.

[115] J.R. Dutcher. Polysaccharides nanoparticles. US12/809,629. 2010.

[116] T. C. Zion, A. Zarur, J.Y. Ying. Stimuli- responsive system for controlled drug delivery. US10/740,436. 2009.

[117] B. R. Conway, D. Gherghel. Chronotherapeutic ocular delivery system comprising a combination of prostaglandin and a beta- blocker for treating primary glaucoma. US11/992,272. 2009.

Advanced Applications of Polysaccharides and their Composites   Materials Research Forum LLC
Materials Research Foundations **73** (2020) 65-85   https://doi.org/10.21741/9781644900772-3

Chapter 3

Polysaccharide Membranes for Drug Delivery System in the Skin Lesion

L.N. Andrade[1], D.M.L. Oliveira[2], C.F. Silva[4], R.G. Amaral[1], M.V. Chaud[5], E. Souto[6], Su Shin Ryon[7], P. Severino[2,3,8*]

[1]Federal Universityof Sergipe, 330 Governador Marcelo Déda Ave, 49400-000 Lagarto, SE, Brazil

[2]Tiradentes University and Institute of Technology and Research, 300 MuriloDantas Ave, 49.032-490 Aracaju, SE, Brazil

[3]Institute of Technology and Research, 300 MuriloDantas Ave, 49.032-490 Aracaju, SE, Brazil

[4] Instituto de Ciências Ambientais, Químicas e Farmacêuticas, Universidade Federal de São Paulo, Diadema, Brazil

[5] Biomaterials and Nanotechnology Laboratory, University of Sorocaba, Sorocaba, SP, Brazil

[6]Faculty ofPharmacy, Universityof Coimbra (FFUC), Pólo das Ciências da Saúde, Azinhaga de Santa Comba, 3000-548 Coimbra, Portugal; REQUIMTE/LAQV, GroupofPharmaceutical Technology, FacultyofPharmacy, Universityof Coimbra, Coimbra, Portugal

[7] Center for Biomedical Engineering, Department of Medicine, Brigham and Women's Hospital, Harvard Medical School, 65 Landsdowne Street, Cambridge, Massachusetts 02139, USA
[8]Tiradentes Institute, 150 Mt Vernon St, Dorchester, MA 02125, USA

*pattseverino@gmail.com

Abstract

This chapter described the importance of polysaccharides-based membranes for drug delivery systems. The most recent advances in the treatment of skin lesions are related to the use of biomaterials with biological activity that contribute to the cicatrization, reducing the time of cure of the renovated tissue. Polysaccharides are the structural material used in the development of these new dressings because it is biocompatibility and biodegradability. Also, the incorporation of the drug in membrane drug delivery showed to be efficient for treatment. However, the search for an ideal dressing is still underway.

Keywords

Membrane, Wound Healing, Polysaccharide, Drug Delivery, Drug

Contents

1.   Introduction

Skin is our largest organ of the human body (16% of the body weight, 2 m^2 in the surface in adults). The skin recovers and protects the body surface, and it has functions: (i) to regulate the loss of water; (ii) to protect the body against friction; (iii) control the blood flow; and; (iv) to participate in the sensory functions (heat, cold, pressure, pain and touch) [1]. With this, the skin is a vital organ formed by three layers: epidermis, dermis, and hypodermis, from the outermost to the deepest respectively.

The interruption in the continuity of the skin causes the wound, which can reach the epithelial tissue, mucous membranes, or organs with anatomical and physiological compromise [2]. And, it occurs every time the tissue loss goes beyond the dermis, and skin healing is classified into three stages. The first stage denominated inflammatory phase begins soon after the injury; the proliferative is responsible for epidermal reconstitution, the so-called re-epithelialization; and in the remodeling phase deposition of neoformed tissue occurs, which contributes to the maturation of scar tissue [3].

The phenomena described refer to the physiological healing process, but there are situations in which there is a decrease in the response of the organism, such as diabetes mellitus and excessive exposure to radiation. There may also be an increase in this response, as in cases of keloid scar or hypertrophic scar [4, 5]. Normal wounds have "stop" signs that interrupt repair when the epithelization is complete, but there are wound that show difficulty to closed, or ineffective, causing excessive scarring. Mechanisms that lead to over replacement are still objects of investigation [1].

A chronic wound is one that failed to progress in the repair phases in an orderly and timely manner and showed no significant progress toward healing in 30 days [6]. The loss of cutaneous integrity cause disequilibrium in the homeostasis in the tissue, being the process of complicated tissue repair, multifactorial and continuous, mechanism of specific attention and cellular extracellular as cytokines and growth factors [7]. The treatments are based on laser therapy, microcurrent technique, the use of the hyperbaric chamber and the vacuum dressing (Barker vacuum), however, they are generally high-cost artifacts and require a considerable (physical and personal) structure for their use [8].

Recent studies have promoted the incorporation of actives into biomaterials, which point to advantages such as acceleration of the granulation process and reepithelialization, controlled release of the active principle, and increased patient comfort and shorter treatment duration [9].

Innovations about materials for repair of cutaneous lesions, polymers are the most used biomaterials in the medical field, being able to constitute biomembranes, biofilms, gels, hydrocolloids, and hydrogels, associating biological activity with the possibility of drug delivery. However, membranes composed by polysaccharides are the most used because of its easy application, which has in their use alone or in an associated way scientifically proven advantages in the optimization of wound healing processes [10].

The ideal membrane provides for the control of bacterial growth, removes devitalized tissue, and stimulates the growth of keratinocytes. Furthermore, the dressing should have low toxicity, rapid action, no irritation or sensitization, no adhesion, and be effective even in the presence of significant exudate [11]. Also, mention adequate water vapor permeability to maintain the environment of wound humid and drugs as an antimicrobial activity to evicting infection.

2.   Biomaterials

Biomaterials are defined as any pharmacologically inert material capable of interacting with a living organism, not inducing adverse reactions at the site of implantation or even systemically. The origin can be natural or synthetic polymers, ceramics, and metals, and it is used for a period, that improves the cicatrization increasing or replacement of the skin [12].

The first characteristic studied of biomaterials is biocompatibility because it's replacing damaged tissue and providing mechanical support with the minimal patient biological response. The evolution of this process has added concepts such as biodegradability, with the ability to be incorporated or absorbed by the host tissue, and, more recently, the idea of biomimetic, looking for materials that actively participate in the recovery process, acting in the specific tissue, with stimulation at the cellular level [13].

There are various types of biomaterials, and the use of membranes is mostly used for skin lesions. Membranes composed by polysaccharides show biological activity that contributes to the healing process, enhancing the cicatrization and appearance of the skin. It acts as a permeable barrier to avoid water loss, protect from the external environment, and favor gas exchange, have been used in the medical and pharmaceutical industries [10, 14, 15].

Polysaccharides are the main material used in the production of these membranes for wound healing because it shows high biocompatibility, biodegradability, and often, physiological activity. Also, they have the property of associating with damaged tissue, promoting healing, and can also be easily mixed with antimicrobial agents to improve their performance on chronic wounds. Some of the dressings already developed have proven effective; however, the ideal membrane is still in progress [16].

2.1 Polysaccharide

Polysaccharides are composed of monosaccharides joined by glycosidic bonds, obtained in plant bio synthesizer animals and by microorganisms production. There is a long linear or branched chains; it is estimated that in nature, 90% of all carbohydrate mass is in the form of these molecules, which in turn perform a complex of functions [17].

Several criteria should be considered when choosing a polysaccharides material for wound healing because its particular properties direct them to a specific application. In this sense, the forms that the chains may assume, the arrangement of the monomeric units, the presence or absence of particular atoms or functional groups, structural rigidity, chain polarity, and polymer molar mass result in subclasses of compounds which may present differentiated behaviors. The chemical structure of the polymer is crucial for the action and influences degradation, toxicity, and biocompatibility of the membrane [18].

Natural polysaccharides have been outstanding because they are obtained from raw materials from renewable sources, such as corn, sugarcane, cellulose, chitin, and others. The choice for these materials has been highlighted in the area of dressings because they can maintain a controlled microenvironment at the lesion site [19].

The characteristics of biocompatibility and biodegradability have their mechanical properties limited, and to preserve their biological artifice, often the costs of production become high. During the formation of the compounds, some of these problems can be minimized, depending on the association of other polymers or the incorporation of a drug [17]

Despite all the advantages, polysaccharides have some technical limitations that make their preparation and their use as a final product difficult; these difficulties include improving the thermal resistance, the mechanical and rheological properties and the rate of degradation of the same [20]. The modifications of these compounds have been studied and from them are developed blends, composites, and nanocomposites to remedy the cited obstacles and to aggregate studies regarding the properties of gas permeability and the costs of obtaining and processing [21].

The natural polymer is the most used to produce membrane, include as chitin and derivates, carrageenan, pectin, hyaluronic acid, cellulose, agarose, alginate, collagen, silk fibroin, as detailed in Table 1 [22].

Table 1. The polysaccharides applied for wound healing of integrity to traumatized tissue.

Polysaccharides	Drug incorporated	Formulation	Application	Pharmacological	Reference
Agarose	Tannicacid	Hydrogel	Wound healing	*In vitro*	[23]
	Chitosan	Hydrogel	Skinregeneration	*In vitro* and *In vivo*	[24]
Alginate	Vicenin-2	Hydrocolloid Film	Wound in diabetics	*In vivo*	[25]
	Aloe vera + Moringa Oleifera	Scaffolds	Wound healing	*In vitro*	[26]
	Protamine and hyaluronan oligosaccharide	Hydrogel	Skin wounds in diabetics	*In vivo*	[27]
	Diclofenac sodium	Hydrogel Thermosensitive	Treatingre-infectedwounds	*In vitro*	[28]
	Tetracycline hydrochloride	Films	Wounddressing	*In vitro*	[29]
	Fattyacids + vitamins A + E	Hydrogel	Wounds in diabetic feet	Study with humans	[30]
	Ciprofloxacin	Films	Healing of infected foot ulcers	*In vitro*	[31]
	Povidone-iodine	Films	Wound healing	*In vivo*	[32]
	Mitsugumin 53	Hydrogel	ChronicWound healing	*In vitro*	[33]
	Insulin	Sponge dressing	Healingofburn wounds	*In vivo*	[34]
Alginate/Hyaluronic acid	Platelet lysate and vancomycin hydrochloride	Gel	Chronicskinulcers	*In vitro* and *Ex vivo*	[35]
Alginate/Pectin	Simvastatin	Films	Wounddressing	*In vivo*	[36]
Carrageenan	Streptomycin Anddiclofenac	Films	Woundhealing	*In vitro*	[37]

Cellulose	Montmorillonite and modified montmorillonites	Artificial skin	Burn skin and tissue Regeneration	*In vitro* and *In vivo*	[38]
	Styelaclava tunics and Broussonetiakazinoki bark	Liquid bandage	Healing of surgical wounds on the skin	*In vivo*	[39]
	Diclofenac	Hydrocolloidmembrane	Wound healing	*In vitro*	[40]
	Sericinand polyhexamethylenebiguanide	Bacterialcellulose	Woundtreatment	*In vivo* and Study with humans	[41]
	Ibuprofen	Membrane	Vasculogenicwounds in diabetics	Study with humans	[42]
Chitosan	Beta-glucan	Gel	Wound healing in diabetic feet	Study with humans	[43]
	Isosorbidedinitrate	Gel	Wound in diabetics	Study with humans	[44]
	Chlorhexidine, allantoin,anddexpanthenol	Topical gel	Postoperative healing by third molar extraction	Study with humans	[45]
	PDGF-BB	Scaffold	Wound healing	*In vitro*	[46]
	Phenytoin	Hydrogel	Cutaneouswounds	*In vitro* and *In vivo*	[47]
	Rosuvastatin	Scaffolds	Skin healing	*In vitro* and *In vivo*	[48]
	-	Membrane	Cutaneouswounds	*In vivo*	[49]
Galactomannan	Cramoll	Films	Topical wounds	*In vitro* and *In vivo*	[22]
	Frutalin	Hydrogelandmembranescaffold	Excision wound repair	*In vitro* and *In vivo*	[50]
Guargum (polymericgalactomannan)	Silver nanoparticles	Nanocomposite	WoundHealing	*In vivo*	[51]
Heparin	Human epidermal growth factor	Hydrogel	Skinwoundhealing	*In vivo*	[52]
	Fibroblastgrowthfactor	hydrogelfilms	Cutaneouswounds in diabetics	*In vivo*	[53]
Hyaluronic acid	Adipose-derived stem cells	Hydrogel	Vascularization for skin wounds and tissue	*In vivo*	[54]

			engineering			
Pectin	Honey	Hydrogel	woundhealing	*In vivo*	[55]
Sericin/Agarose	Lysozyme	Gel	Woundhealing	*In vitro*	[56]
Xanthan and Galactomannan	Curcumin	Hydrogel	woundhealing	*In vivo*	[57]
Xanthan and Chitosan	Multipotent mesenchymal stromal cells	Scaffolds	Healingcutane ouswounds	*In vivo*	[58]

Generally, the membrane composed by polysaccharides is not-toxic, biocompatible, and appropriate for metabolism and excretion by the physiological pathways [59]. And, the membrane is used as a drug delivery system that drug can be chemically bound, dispersed or dissolved in the polymer structure. The most used method for the production of membranes is called solvent evaporation (casting) and consists of solubilizing the polymer in a suitable solvent, pouring it onto a smooth surface and drying, in the exhaust hood, stove with air circulation or ambient temperature, it to the formation of the membrane [15].

Membranes composed by chitosan and alginate may be dense or porous, produced to meet the final application. However, films based on this blend have limitations on mechanical properties [60]. Chitosan stands out as a raw material for controlled drug release systems due to all its characteristics that involve biodegradability and biocompatibility, but the main highlight is the devices developed for a cutaneous application, as it presents itself as a bioactive polymer, accelerating the wound healing and the synthesis of collagen by fibroblasts in the initial healing phase [61].

It is expected that the controlled release of drugs from the membrane will be in an area in continuous evolution, aiming the development of systems that allow releasing drugs in a determined place, controlled speed, and in a specific time [18]. According to [11] the types of agents used in the treatment of the lesion generally follows a process: removal of residues present in the injury, retention of liquids found, promotion of heat in the wound bed, chemical-physical protection of the region and cover, destruction of areas deaths among others. The injured skin usually needs to be covered by a bandage to minimize the loss of its functions.

Membranes with the incorporation of drugs are more effective in several aspects, since improving the healing process of the skin and facilitate the treatment of the patient. Exist various types of molecules that can be incorporated into dressings, for example, agents antimicrobial, analgesics, anti-inflammatories, anesthetics, and anti-allergies, among others, depending on the intended use of the dressing. Control of the release of the agent contained in the dressing is essential as it improves the conventional efficacy in the local of application.

The controlled release polymeric systems represent an effective strategy for the incorporation of drugs, and some advances in this modality are described in the literature:

- (i)   control of the release of the drug;
- (ii)   reduction of toxic and sub-therapeutic doses;
- (iii)   monitoring drug levels at the site of application;
- (iv)   obtaining high drug concentrations at the wound healing and;
- (v)   targeting specific

It decreases the dosage interval and reduces unwanted side effects since it uses a smaller amount of the active principle, resulting in the lower cost [62]. The possibility of allying the properties of base polymers, such as adhesion, to innovations and that offers protection and control in the release process, constituting excellent strategies to increase the bioavailability of drugs. Among the various drugs already incorporated into membranes reported in the literature are α-tocopherol acetate [63], Curcumin [64], propolis [65], vitamin C [66], ciprofloxacin [67].

3.   Application skin drug delivery

The intention to restore the integrity of traumatized tissue, from 3500 to 2500 BC, Egyptian topical agent test writings or cover types based on honey, grease, flax, and various kinds of feces are written. With over the centuries, knowledge of the pathophysiological basis of tissue repair has advanced and mobilized researchers and industries to develop increasingly effective and appropriate dressings for each type of wound and to accelerate or allow healing (Wu et al., 2017).

In this sense, polysaccharide-based membranes for drug administration have been a delivery system that can effectively enable their full therapeutic benefit (Yassue-Cordeiro et al., 2019). That is, the intended purpose of a delivery system, in the context of wound healing, must have its bioactivity and bioavailability, preventing rapid dilution in wound fluid and systemic uptake and distribution, as well as maintaining its release into the wound applied site by determining a physiologically relevant rate and duration. Thus, by achieving these goals, a successful delivery system will also minimize the dosing frequency and application required for efficacy (Johnson et al., 2015). In this sense, different bioactive substances have been incorporated into dressings aiming at the prevention and control of bacterial infections, thus improving speed in the healing process (Rao et al., 2016, Song et al., 2018). Among many, we can mention the incorporation of antimicrobial agents such as silver (Meng et al., 2010) and bacitracin (Khan et al., 2015), of analgesics and antipyretics such as paracetamol (Fernández-Hervás and Fell, 1998), and anti-inflammatory agents such as sodium diclofenac (Gul et al., 2018).

The inclusion of antibiotics in dressings is of high relevance for the treatment of skin lesions, to promote the control of infectious processes at the site of injury. These drugs have been widely used by the pharmaceutical industry as an active ingredient in ointments, whether or not associated with other antibiotics (Kim et al., 2017). Depending on the solubility of these drugs in aqueous media will enable their incorporation into natural polymer-based devices by swelling these matrices in aqueous solutions containing the drug. Thus, the inclusion of antibiotics in membranes is very relevant, as it would allow not only the recovery of tissues injured by the action of biopolymers but also the growth control of undesirable microorganisms (Kumar et al., 2018)

Recently, Sequeira et al., 2019 [68] developed a nanofibrous membrane composed of poly(vinyl alcohol) (PVA), anti-inflammatory, and antibacterial actives. The drug delivery system was primordial for the success of wound healing because Lavanda oil (antibacterial) showed an initial burst release, and the ibuprofen (anti-inflammatory) released in the inflammatory phase. Also, no-toxic effect was observed in fibroblast cells, and the antibacterial activity was relevant for gram-positive and negative bacterial strains.

Following this same line Garms et al., 2019 [69] developed a membrane using natural rubber latex and incorporated moxifloxacin as antimicrobial active to drug delivery. It is known that the previously used natural rubber latex improves angiogenesis and facilitate the wound healing. The mechanical properties did not change with the incorporation of moxifloxacin, and cell viability showed the biomaterial was biocompatibility.

The incorporation of silver nanoparticle as antimicrobial activity in the membrane is widely published in the literature [70-72]. Among the works Tarusha et al., 2019 [73] developed a mixture of polymers composed of alginate, hyaluronic acid, and Chitlac-silver nanoparticles to treat chronic wound healing. The *in vitro* biological assays the silver nanoparticle was sufficient to promote antibacterial activity, and the hyaluronic acid was responsible for improving the healing. Agreeing with this, Kumar et al., 2018 [74] details that polysaccharide-based membrane systems represent a promising alternative for the development and production of bioactive dermal dressings. And, the complexation of natural polymers showed excellent performance when applied as dermal dressings, presenting adequate physicochemical and biological characteristics for this purpose. Ideally, however, a bandage should not only protect the injury but also promote the healing process.

4.   Marketing

Naseri-Nosar et al., 2018 [75] demonstrate different forms of dressings based on polysaccharides, as well as their combination with other natural polymers, dressings containing metals, antibiotics, proteins, and other bioactive substances to accelerate healing. Most of the research was based on bacterial cellulose and chitosan and lacking those who mentioned antimicrobial activity are commercially available, as shown in Table 2.

In this growing market for biomaterials, polysaccharides appear often due to its chemical structure, allied to attractive properties such as none toxicity, hydrophilicity, biocompatibility, biodegradability and can be successfully used as raw materials in production of an extensive range of biomedical devices However, the cost is very high, and in developing countries, the people do not have access to this kind of treatment [76].

Table 2. Examples of dressings consisting of polysaccharides commercially available.

Polysaccharides	Commercial name of dressing	Manufacturer	Applications
Alginate	Algicell™	Derma Sciences	Ulcers (diabetic, leg, and pressure) and wounds (traumatic and surgical).
	Alginateand CMC™	ReliaMed	Ulcers (pressure and legs) and skin wounds (deep and superficial).
	Algisite M™	Smith &Nephew	Ulcers (pressure, legs, and diabetic feet), skin wounds (deep, superficial, and surgical).
	Algosteril™	Systagenix	Ulcers (pressure, legs, and diabetics), wounds (hemorrhagic and infected), pilonidal abscess, amputation site, and skin graft.
	Biataim™	Coloplast	Ulcers (pressure, venous leg, arterial and diabetic) and wounds caused by trauma.
	Comfeel™Plus	Coloplast	Ulcers (pressure and legs), wounds (superficial and postoperative), burns, and transplanted areas.
	Curasorb™	Kendall Healthcare	Wound (Deep, superficial and with exudate)
	CurasorbZinc ™	Covidien/Medtronic	Ulcers (venous, arterial, diabetic and pressure), Wound (Deep, superficial and with moderate to high exudate), second-degree burns

			and transplant areas.
	Gentell™ CalciumAlgin ate	Gentell	Ulcers (leg, diabetic foot, and pressure), Wounds (superficial and surgical) and minor burns.
	Gentell ™ CalciumAlgin ate Ag (Silver)	Gentell	Infected wounds Including all wounds mentioned in the previous section.
	Kalginate™	DeRoyal	Ulcers (Venous, diabetic and pressure), Wounds (Surgical, infected) and second-degree burns.
	Kaltostat	ConvaTec	Ulcers (venous, arterial, diabetic and pressure), wounds (surgical, traumatic and superficial), first degree burns, and transplanted areas.
	Maxorb™ II	Medline Industries	Ulcers (leg, diabetic and pressure), Wounds (superficial, deep and postoperative), first and second degree burns, transplanted areas, and graft area.
	McKesson™ CalciumAlgin ateDressing	McKesson	Wound exudate absorption, minor bleeding control, and filling dead space in wounds.
	Melgisorb™ Plus	Molnlycke Health Care	Ulcers (arterial, venous, pressure and diabetic), surgical wounds, and transplanted areas.
	Sorbalgon ™	Hartmann USA, inc.	For use with chronic or acute moderate to slowly draining wounds.
	Sorbsan™	Aspen	Ulcers (in the legs - arterial or venous, pressure and diabetic), wounds (surgical and traumatic), transplanted and graft areas.
	Tegagen™	3M Healthcare	Dermal ulcers: venous, arterial, diabetic and pressure ulcers; surgical dehiscences; traumatic wounds and abrasions; neoplastic lesions; in deep wounds, it is used to fill dead spaces
	Tegaderm™	3M	Ulcers (arterial, venous and diabetic), wounds (surgical and traumatic), transplanted areas and control of minor bleeding.
Alginate/carboxym ethylcellulose	Maxorb Extra™	Medline Industries	Ulcers (pressure, leg, and diabetic), wounds (superficial, deep and post-surgical), 1st and 2nd degree burns and transplanted areas.
	SeaSorb™ Soft	Coloplast	Wounds of moderate to high exudation and as support in the hemostasis of minor bleeds in

			superficial wounds are also indicated for venous leg ulcers; Pressure ulcers; Diabetic foot ulcers; Skin donor areas; 2nd degree burns with intense exudation, and For filling wound cavities in general
	Tegaderm™ Alginate Ag Silver	3M	Ulcers (pressure, leg, and diabetic), wounds (superficial and post-surgical trauma), 2nd-degree burns, transplanted areas, graft site and can be used under compression bandage
Chitin (N-acetylglucosamine)	Syvek™ NT Patch	Marine Polymer Tech	Bleeding wounds (vascular access site) and bleeding control (Patients on hemodialysis or anticoagulant treatment).
Chitosan	HemConBandage™	HemCon	Hospital, hemostatic and antibacterial dressing
Pectin/carboxylmethylcellulose	Combiderm™	ConvaTecLtd.	Chronic exudative ulcers and acute wounds: bruises, lacerations, biopsies, and open or closed surgical wounds.
	Duoderm™	ConvaTecLtd.	Skin lesions that are superficial, dry, or slightly exudative; Post-surgical wounds; Prevention of skin lesions; Pressure injuries in stages I and II.
	Granuflex™	ConvaTecLtd.	Ulcers (pressure and leg) Wounds (acute and chronic), burns, transplant area, and skin graft.
	GranugelR paste	ConvatecLtd.	Ulcers (by pressure and in the legs) and Skin sores.

Conclusion

Almost all people suffer from some kind of wounds from surgery, disease like diabetes, cuts, and burns in their life. To cover the wounded skin is attractive to protect and, facilitate the cicatrization. Also, antimicrobial dressing is an alternative to treat the infectious wound as active drugs (antibiotics), nanoparticles, or natural products. However, further improvements in this type of dressings are required; suggesting the co-administration of antibacterial agents is expected to lead to increasing therapeutic results. Besides, dressings containing sensors and therapeutic molecules may also be produced to perform the monitoring and treatment of an infected wound. The increasing worldwide investments in innovative technologies and techniques, the development and discovery of new wound care products continues to be an essential component of scientific research to apply in medical clinics.

References

[1]   H. Sorg, D.J. Tilkorn, S. Hager, J. Hauser, U.J. Mirastschijski, Skin wound healing: an update on the current knowledge and concepts, Eur. Surg. Res. 58 (2017) 81-94. https://doi.org/10.1159/000454919

[2]   Z.D. Draelos, Skin lightening preparations and the hydroquinone controversy, Dermatol. Ther. 20 (2007) 308-313. https://doi.org/10.1111/j.1529-8019.2007.00144.x

[3]   B. Horst, G. Chouhan, N.S. Moiemen, L.M, Advances in keratinocyte delivery in burn wound care, Adv. Drug Deliv. Rev. 123 (2018) 18-32. https://doi.org/10.1016/j.addr.2017.06.012

[4]   K. Järbrink, G. Ni, H. Sönnergren, A. Schmidtchen, C. Pang, R. Bajpai, J.J., Prevalence and incidence of chronic wounds and related complications: a protocol for a systematic review, Syst. Rev. 5 (2016) 152. https://doi.org/10.1186/s13643-016-0329-y

[5]   I. Garcia-Orue, G. Gainza, F.B. Gutierrez, J.J. Aguirre, C. Evora, J.L. Pedraz, R.M. Hernandez, A. Delgado, M. Igartua, Novel nanofibrous dressings containing rhEGF and Aloe vera for wound healing applications, Int. J. Pharm. 523 (2017) 556-566. https://doi.org/10.1016/j.ijpharm.2016.11.006

[6]   R. Capoano, R. Businaro, B. Kolce, A. Biancucci, S. Izzo, L. De Felice, B. Salvati, Multidisciplinary Approaches to the Stimulation of Wound Healing and Use of Dermal Substitutes in Chronic Phlebostatic Ulcers, Wound Healing-Current Perspectives, IntechOpen2019. https://doi.org/10.5772/intechopen.81791

[7]   M. Schrementi, L. Chen, L.A. DiPietro, The importance of targeting inflammation in skin regeneration, Skin Tissue Models, Elsevier 2018, pp. 255-275. https://doi.org/10.1016/B978-0-12-810545-0.00011-5

[8]   J. Davidson, Current concepts in wound management and wound healing products, Vet Clin North Am Small Anim. Pract. 45 (2015) 537-564. https://doi.org/10.1016/j.cvsm.2015.01.009

[9]   H.S. Debone, P.S. Lopes, P. Severino, C.M.P. Yoshida, E.B. Souto, C.F. da Silva, Chitosan/Copaiba oleoresin films for would dressing application, Int. J. Pharm. 555 (2019) 146-152. https://doi.org/10.1016/j.ijpharm.2018.11.054

[10] P.B. Milan, S. Kargozar, M.T. Joghataie, A. Samadikuchaksaraei, Nanoengineered biomaterials for skin regeneration, Nanoengineered Biomaterials for Regenerative Medicine, Elsevier 2019, pp. 265-283. https://doi.org/10.1016/B978-0-12-813355-2.00011-9

[11] M. Naseri-Nosar, Z.M.J.C.p. Ziora, Wound dressings from naturally-occurring polymers: A review on homopolysaccharide-based composites, Carbohydr. Polym. 189 (2018) 379-398. https://doi.org/10.1016/j.carbpol.2018.02.003

[12] A. Sandak, J. Sandak, M. Brzezicki, A. Kutnar, Biomaterials for Building Skins, Bio-based Building Skin, Springer 2019, pp. 27-64. https://doi.org/10.1007/978-981-13-3747-5_2

[13] M. Niinomi, Design and development of metallic biomaterials with biological and mechanical biocompatibility, J. Biomed. Mater. Res. A. 107 (2019) 944-954. https://doi.org/10.1002/jbm.a.36667

[14] D. Rodrigues, A.C. Viotto, R. Checchia, A. Gomide, D. Severino, R. Itri, M.S. Baptista, W.K. Martins, Mechanism of Aloe Vera extract protection against UVA: shelter of lysosomal membrane avoids photodamage, Photochem. Photobiol. Sci. 15 (2016) 334-350. https://doi.org/10.1039/C5PP00409H

[15] P. Severino, A. Feitosa, I. Lima-Verde, M. Chaud, C. Da Silva, R. De Lima, R. Amaral, L. Andrade, Production and Characterization of Mucoadhesive Membranes for Anesthetic Vehiculation, Chem. Eng. Transactions. 64 (2018) 421-426.

[16] P. Balakrishnan, V. Geethamma, M.S. Sreekala, S. Thomas, Polymeric biomaterials: State-of-the-art and new challenges, Fundamental Biomaterials: Polymers, Elsevier 2018, pp. 1-20. https://doi.org/10.1016/B978-0-08-102194-1.00001-3

[17] J.H. Shepherd, D. Howard, A.K. Waller, H.R. Foster, A. Mueller, T. Moreau, A.L. Evans, M. Arumugam, G.B. Chalon, E.J.B. Vriend, Structurally graduated collagen scaffolds applied to the ex vivo generation of platelets from human pluripotent stem cell-derived megakaryocytes: Enhancing production and purity, J. Biomaterials. 182 (2018) 135-144. https://doi.org/10.1016/j.biomaterials.2018.08.019

[18] N. Kamaly, B. Yameen, J. Wu, O. Farokhzad, Degradable controlled-release polymers and polymeric nanoparticles: mechanisms of controlling drug release, Chem. Rev. 116 (2016) 2602-2663. https://doi.org/10.1021/acs.chemrev.5b00346

[19] T.A. Saleh, V.K. Gupta, Nanomaterial And Polymer Membranes: Synthesis, Characterization, And Applications, Elsevier 2016.

[20] H. Hamedi, S. Moradi, S.M. Hudson, A. Tonelli, Chitosan based hydrogels and their applications for drug delivery in wound dressings: A review, Carbohydr. Polym. 199 (2018) 445-460. https://doi.org/10.1016/j.carbpol.2018.06.114

[21] B. Joseph, R. Augustine, N. Kalarikkal, S. Thomas, B. Seantier, Y. Grohens, Recent advances in electrospun polycaprolactone based scaffolds for wound healing and skin bioengineering applications, Mater Today Commun. 19 (2019) 319-335. https://doi.org/10.1016/j.mtcomm.2019.02.009

[22] P.B. Albuquerque, P.A. Soares, A.C. Aragão-Neto, G.S. Albuquerque, L.C. Silva, M.H. Lima-Ribeiro, J.C.S. Neto, L.C. Coelho, M.T. Correia, J.A. Teixeira, Healing activity evaluation of the galactomannan film obtained from Cassia grandis seeds with immobilized Cratylia mollis seed lectin, Int. J. Biol. Macromol. 102 (2017) 749-757. https://doi.org/10.1016/j.ijbiomac.2017.04.064

[23] N. Ninan, A. Forget, V.P. Shastri, N.H. Voelcker, A. Blencowe, Antibacterial and anti-inflammatory pH-responsive tannic acid-carboxylated agarose composite hydrogels for wound healing, ACS Appl. Mater. Interfaces. 8 (2016) 28511-28521. https://doi.org/10.1021/acsami.6b10491

[24] S.P. Miguel, M.P. Ribeiro, H. Brancal, P. Coutinho, I. Correia, Thermoresponsive chitosan–agarose hydrogel for skin regeneration, Carbohydr. Polym. 111 (2014) 366-373. https://doi.org/10.1016/j.carbpol.2014.04.093

[25] W.S. Tan, P. Arulselvan, S.-F. Ng, C.N.M. Taib, M.N. Sarian, S. Fakurazi, Improvement of diabetic wound healing by topical application of Vicenin-2 hydrocolloid film on Sprague Dawley rats, BMC Complement Altern. Med. 19 (2019) 20. https://doi.org/10.1186/s12906-018-2427-y

[26] I. Rubio-Elizalde, J. Bernáldez-Sarabia, A. Moreno-Ulloa, C. Vilanova, P. Juárez, A. Licea-Navarro, A. Castro-Ceseña, Scaffolds based on alginate-PEG methyl ether methacrylate-Moringa oleifera-Aloe vera for wound healing applications, Carbohydr. Polym. 206 (2019) 455-467. https://doi.org/10.1016/j.carbpol.2018.11.027

[27] T. Wang, Y. Zheng, Y. Shi, L. Zhao, pH-responsive calcium alginate hydrogel laden with protamine nanoparticles and hyaluronan oligosaccharide promotes diabetic wound healing by enhancing angiogenesis and antibacterial activity, Drug. Deliv. Transl. Res. 9 (2019) 227-239. https://doi.org/10.1007/s13346-018-00609-8

[28] A. Montaser, M. Rehan, M.E. El-Naggar, pH-Thermosensitive hydrogel based on polyvinyl alcohol/sodium alginate/N-isopropyl acrylamide composite for treating re-

infected wounds, Int. J. Biol. Macromol. 124 (2019) 1016-1024.
https://doi.org/10.1016/j.ijbiomac.2018.11.252

[29] H. Türe, Characterization of hydroxyapatite-containing alginate–gelatin composite films as a potential wound dressing, Int. J. Biol. Macromol. 123 (2019) 878-888. https://doi.org/10.1016/j.ijbiomac.2018.11.143

[30] M.A.G. Barbosa, A.O. Paggiaro, V.F. de Carvalho, C. Isaac, R. Gemperli, Effects of hydrogel with enriched sodium alginate in wounds of diabetic patients, Plast. Surg. Nurs. 38 (2018) 133-138. https://doi.org/10.1097/PSN.0000000000000228

[31] A. Ahmed, J. Boateng, Calcium alginate-based antimicrobial film dressings for potential healing of infected foot ulcers, Ther. Deliv. 9 (2018) 185-204. https://doi.org/10.4155/tde-2017-0104

[32] M. Summa, D. Russo, I. Penna, N. Margaroli, I.S. Bayer, T. Bandiera, A. Athanassiou, R. Bertorelli, A biocompatible sodium alginate/povidone iodine film enhances wound healing, Eur. J. Pharm. Biopharm. 122 (2018) 17-24. https://doi.org/10.1016/j.ejpb.2017.10.004

[33] M. Li, H. Li, X. Li, H. Zhu, Z. Xu, L. Liu, J. Ma, M. Zhang, A bioinspired alginate-gum arabic hydrogel with micro-/nanoscale structures for controlled drug release in chronic wound healing, ACS Appl. Mater. Interfaces. 9 (2017) 22160-22175. https://doi.org/10.1021/acsami.7b04428

[34] S. Dhall, J.P. Silva, Y. Liu, M. Hrynyk, M. Garcia, A. Chan, J. Lyubovitsky, R.J. Neufeld, M. Martins-Green, Release of insulin from PLGA–alginate dressing stimulates regenerative healing of burn wounds in rats, Clin. Sci. (Lond). 129 (2015) 1115-1129. https://doi.org/10.1042/CS20150393

[35] S. Rossi, M. Mori, B. Vigani, M. Bonferoni, G. Sandri, F. Riva, C. Caramella, F. Ferrari, A novel dressing for the combined delivery of platelet lysate and vancomycin hydrochloride to chronic skin ulcers: Hyaluronic acid particles in alginate matrices, Eur. J. Pharm. Sci. 118 (2018) 87-95. https://doi.org/10.1016/j.ejps.2018.03.024

[36] M. Rezvanian, M.C.I.M. Amin, S.-F. Ng, Development and physicochemical characterization of alginate composite film loaded with simvastatin as a potential wound dressing, Carbohydr. Polym. 137 (2016) 295-304. https://doi.org/10.1016/j.carbpol.2015.10.091

[37] J.S. Boateng, H.V. Pawar, J. Tetteh, Polyox and carrageenan based composite film dressing containing anti-microbial and anti-inflammatory drugs for effective wound

healing,Int. J. Pharm. 441 (2013) 181-191.
https://doi.org/10.1016/j.ijpharm.2012.11.045

[38] W. Sajjad, T. Khan, M. Ul-Islam, R. Khan, Z. Hussain, A. Khalid, F. Wahid, Development of modified montmorillonite-bacterial cellulose nanocomposites as a novel substitute for burn skin and tissue regeneration, Carbohydr. Polym. 206 (2019) 548-556. https://doi.org/10.1016/j.carbpol.2018.11.023

[39] J.J. Park, J.E. Kim, W.B. Yun, M.R. Lee, J.Y. Choi, B.R. Song, H.J. Son, Y. Lim, H.G. Kang, B.S. An, Therapeutic effects of a liquid bandage prepared with cellulose powders from Styelaáclava tunics and Broussonetiaákazinoki bark: Healing of surgical wounds on the skin of Sprague Dawley rats, Mol. Med. Rep. 19 (2019) 452-460. https://doi.org/10.3892/mmr.2018.9668

[40] T. Maver, L. Gradišnik, D.M. Smrke, K.S. Kleinschek, U. Maver, Systematic evaluation of a diclofenac-loaded carboxymethyl cellulose-based wound dressing and its release performance with changing ph and temperature, AAPS Pharm Sci. Tech. 20 (2019) 29. https://doi.org/10.1208/s12249-018-1236-4

[41] S. Napavichayanun, S. Ampawong, T. Harnsilpong, A. Angspatt, P. Aramwit, Inflammatory reaction, clinical efficacy, and safety of bacterial cellulose wound dressing containing silk sericin and polyhexamethylene biguanide for wound treatment, Arch. Dermatol. Res. 310 (2018) 795-805. https://doi.org/10.1007/s00403-018-1871-3

[42] M.D. Umebayashi Zanoti, H. Megumi Sonobe, S.J. Lima Ribeiro, A.M. Minarelli Gaspar, Development of coverage and its evaluation in the treatment of chronic wounds, Invest. Educ. Enferm. 35 (2017) 330-339. https://doi.org/10.17533/udea.iee.v35n3a09

[43] K. Cutting, The cost-effectiveness of a novel soluble beta-glucan gel, J. Wound Care. 26 (2017) 228-234. https://doi.org/10.12968/jowc.2017.26.5.228

[44] S.E. Totsuka Sutto, Y.I. Rodríguez Roldan, E.G. Cardona Muñoz, T.A. Garcia Cobian, S. Pascoe Gonzalez, A. Martinez Rizo, M. Mendez del Villar, L. Garcia Benavides, Efficacy and safety of the combination of isosorbide dinitrate spray and chitosan gel for the treatment of diabetic foot ulcers: A double-blind, randomized, clinical trial, Diab. Vasc. Dis. Res. 15 (2018) 348-351. https://doi.org/10.1177/1479164118769528

[45] M. Madrazo-Jiménez, Á. Rodríguez-Caballero, M.-Á. Serrera-Figallo, R. Garrido-Serrano, A. Gutiérrez-Corrales, J.-L. Gutiérrez-Pérez, D. Torres-Lagares, The effects of a topical gel containing chitosan, 0, 2% chlorhexidine, allantoin and

despanthenol on the wound healing process subsequent to impacted lower third molar extraction, Med. Oral. Patol Oral Cir. Bucal. 21 (2016) e696. https://doi.org/10.4317/medoral.21281

[46] M. Piran, S. Vakilian, M. Piran, A. Mohammadi-Sangcheshmeh, S. Hosseinzadeh, A. Ardeshirylajimi, In vitro fibroblast migration by sustained release of PDGF-BB loaded in chitosan nanoparticles incorporated in electrospun nanofibers for wound dressing applications, Artif Cells Nanomed. Biotechnol. 46 (2018) 511-520. https://doi.org/10.1080/21691401.2018.1430698

[47] A.M. Cardoso, E.G. de Oliveira, K. Coradini, F.A. Bruinsmann, T. Aguirre, R. Lorenzoni, R.C.S. Barcelos, K. Roversi, D.R. Rossato, A.R. Pohlmann, Chitosan hydrogels containing nanoencapsulated phenytoin for cutaneous use: Skin permeation/penetration and efficacy in wound healing, Mater. Sci. Eng. C Mater Biol Appl. 96 (2019) 205-217. https://doi.org/10.1016/j.msec.2018.11.013

[48] A. Maged, A.A. Abdelkhalek, A.A. Mahmoud, S. Salah, M.M. Ammar, M.M. Ghorab, Mesenchymal stem cells associated with chitosan scaffolds loaded with rosuvastatin to improve wound healing, Eur. J. Pharm. Sci. 127 (2019) 185-198. https://doi.org/10.1016/j.ejps.2018.11.002

[49] P. Nordback, S. Miettinen, M. Kääriäinen, A. Haaparanta, M. Kellomäki, H. Kuokkanen, R. Seppänen, Chitosan membranes in a rat model of full-thickness cutaneous wounds: Healing and IL-4 levels, J. Wound Care. 24 (2015) 245-251. https://doi.org/10.12968/jowc.2015.24.6.245

[50] F.D. de Sousa, P.D. Vasconselos, A.F.B. da Silva, E.F. Mota, A. da Rocha Tomé, F.R. da Silva Mendes, A.M.M. Gomes, D.J. Abraham, X. Shiwen, J.S. Owen, Hydrogel and membrane scaffold formulations of Frutalin (breadfruit lectin) within a polysaccharide galactomannan matrix have potential for wound healing, Int. J. Biol. Macromol. 121 (2019) 429-442. https://doi.org/10.1016/j.ijbiomac.2018.10.050

[51] R. Ghosh Auddy, M.F. Abdullah, S. Das, P. Roy, S. Datta, A. Mukherjee, New guar biopolymer silver nanocomposites for wound healing applications, Biomed. Res. Int. 2013 (2013). https://doi.org/10.1155/2013/912458

[52] M. Goh, Y. Hwang, G. Tae, Epidermal growth factor loaded heparin-based hydrogel sheet for skin wound healing, Carbohydr. Polym. 147 (2016) 251-260. https://doi.org/10.1016/j.carbpol.2016.03.072

[53] Y. Liu, S. Cai, X.Z. Shu, J. Shelby, G.D. Prestwich, Release of basic fibroblast growth factor from a crosslinked glycosaminoglycan hydrogel promotes wound

healing, Wound. Repair. Regen. 15 (2007) 245-251. https://doi.org/10.1111/j.1524-475X.2007.00211.x

[54] G. Eke, N. Mangir, N. Hasirci, S. MacNeil, V. Hasirci, Development of a UV crosslinked biodegradable hydrogel containing adipose derived stem cells to promote vascularization for skin wounds and tissue engineering, Biomaterials. 129 (2017) 188-198. https://doi.org/10.1016/j.biomaterials.2017.03.021

[55] G. Giusto, C. Vercelli, F. Comino, V. Caramello, M. Tursi, M. Gandini, A new, easy-to-make pectin-honey hydrogel enhances wound healing in rats, BMC Complement Altern. Med. 17 (2017) 266. https://doi.org/10.1186/s12906-017-1769-1

[56] M. Yang, Y. Wang, G. Tao, R. Cai, P. Wang, L. Liu, L. Ai, H. Zuo, P. Zhao, A. Umar, Fabrication of sericin/agrose gel loaded lysozyme and its potential in wound dressing application, Nanomaterials. 8 (2018) 235. https://doi.org/10.3390/nano8040235

[57] H.S. Koop, R.A. de Freitas, M.M. de Souza, R. Savi-Jr, J.L.M. Silveira, Topical curcumin-loaded hydrogels obtained using galactomannan from Schizolobium parahybae and xanthan, Carbohydr. Polym. 116 (2015) 229-236. https://doi.org/10.1016/j.carbpol.2014.07.043

[58] M.Z. Bellini, C. Caliari-Oliveira, A. Mizukami, K. Swiech, D.T. Covas, E.A. Donadi, P. Oliva-Neto, A.M. Moraes, Combining xanthan and chitosan membranes to multipotent mesenchymal stromal cells as bioactive dressings for dermo-epidermal wounds, J. Biomater. Appl. 29 (2015) 1155-1166. https://doi.org/10.1177/0885328214553959

[59] L. Upadhyaya, M. Semsarilar, D. Quémener, R. Fernández-Pacheco, G. Martinez, R. Mallada, I.M. Coelhoso, C.A. Portugal, J.G. Crespo, Block copolymer based novel magnetic mixed matrix membranes-magnetic modulation of water permeation by irreversible structural changes, J. Memb. Sci. 551 (2018) 273-282. https://doi.org/10.1016/j.memsci.2018.01.032

[60] D.R. Goudie, M. D'Alessandro, B. Merriman, H. Lee, I. Szeverényi, S. Avery, B. D O'Connor, S.F. Nelson, S.E. Coats, A. Stewart, Multiple self-healing squamous epithelioma is caused by a disease-specific spectrum of mutations in TGFBR1, Nat. Genet. 43 (2011) 365. https://doi.org/10.1038/ng.780

[61] X. Li, K. Nan, S. Shi, H. Chen, Preparation and characterization of nano-hydroxyapatite/chitosan cross-linking composite membrane intended for tissue

engineering, Int. J. Biol. Macromol. 50 (2012) 43-49.
https://doi.org/10.1016/j.ijbiomac.2011.09.021

[62] I. Garcia-Orue, J.L. Pedraz, R.M. Hernandez, M. Igartua, Nanotechnology-based delivery systems to release growth factors and other endogenous molecules for chronic wound healing, J. Drug Deliv Sci. Technol. 42 (2017) 2-17. https://doi.org/10.1016/j.jddst.2017.03.002

[63] S. Zahid, H. Khalid, F. Ikram, H. Iqbal, M. Samie, L. Shahzadi, A.T. Shah, M. Yar, A.A. Chaudhry, S.J. Awan, Bi-layered α-tocopherol acetate loaded membranes for potential wound healing and skin regeneration, Mater. Sci. Eng. C Mater. Biol. Appl. 101 (2019) 438-447. https://doi.org/10.1016/j.msec.2019.03.080

[64] X. Liu, L. You, S. Tarafder, L. Zou, Z. Fang, J. Chen, C.H. Lee, Q. Zhang, Curcumin-releasing chitosan/aloe membrane for skin regeneration, Chem. Eng. J. 359 (2019) 1111-1119. https://doi.org/10.1016/j.cej.2018.11.073

[65] T. Krupp, B.D. dos Santos, L.A. Gama, J.R. Silva, W.W. Arrais-Silva, N.C. de Souza, M.F. Américo, P.C. de Souza Souto, Natural rubber–propolis membranes improves wound healing in second-degree burning model, Int. J. Biol. Macromol. 131 (2019) 980-988. https://doi.org/10.1016/j.ijbiomac.2019.03.147

[66] A. Madni, R. Khan, M. Ikram, S.S. Naz, T. Khan, F. Wahid, Fabrication and characterization of chitosan–vitamin c–lactic acid composite membrane for potential skin tissue engineering, Int. J Polymer Sci. 2019 (2019). https://doi.org/10.1155/2019/4362395

[67] G. Ajmal, G.V. Bonde, S. Thokala, P. Mittal, G. Khan, J. Singh, V.K. Pandey, B. Mishra, Ciprofloxacin HCl and quercetin functionalized electrospun nanofiber membrane: fabrication and its evaluation in full thickness wound healing, Artif Cells Nanomed Biotechnol. 47 (2019) 228-240. https://doi.org/10.1080/21691401.2018.1548475

[68] R.S. Sequeira, S.P. Miguel, C.S.D. Cabral, A.F. Moreira, P. Ferreira, I.J. Correia, Development of a poly(vinyl alcohol)/lysine electrospun membrane-based drug delivery system for improved skin regeneration, Int. J. Pharm. 570 (2019) 118640. https://doi.org/10.1016/j.ijpharm.2019.118640

[69] B.C. Garms, F.A. Borges, N.R. de Barros, M.Y. Marcelino, M.N. Leite, M.C. Del Arco, S.L. de Souza Salvador, G.S. Pegorin, K.S.M. Oliveira, M.A.C. Frade, R.D. Herculano, Novel polymeric dressing to the treatment of infected chronic wound, Appl. Microbiol Biotechnol. 103 (2019) 4767-4778. https://doi.org/10.1007/s00253-019-09699-x

[70] X. Li, Z. Cai, D.U. Ahn, X. Huang, Development of an antibacterial nanobiomaterial for wound-care based on the absorption of AgNPs on the eggshell membrane, Colloids Surf. B Biointerfaces. 183 (2019) 110449. https://doi.org/10.1016/j.colsurfb.2019.110449

[71] J. Shao, B. Wang, J. Li, J.A. Jansen, X.F. Walboomers, F. Yang, Antibacterial effect and wound healing ability of silver nanoparticles incorporation into chitosan-based nanofibrous membranes, Mater. Sci. Eng C Mater Biol App. l98 (2019) 1053-1063. https://doi.org/10.1016/j.msec.2019.01.073

[72] S. Pal, R. Nisi, M. Stoppa, A. Licciulli, Silver-Functionalized Bacterial Cellulose as Antibacterial Membrane for Wound-Healing Applications, ACS omega. 2 (2017) 3632-3639. https://doi.org/10.1021/acsomega.7b00442

[73] L. Tarusha, S. Paoletti, A. Travan, E. Marsich, Alginate membranes loaded with hyaluronic acid and silver nanoparticles to foster tissue healing and to control bacterial contamination of non-healing wounds, J. Mater. Sci. Mater Med. 29 (2018) 22. https://doi.org/10.1007/s10856-018-6027-7

[74] S.S.D. Kumar, N.K. Rajendran, N.N. Houreld, H. Abrahamse, Recent advances on silver nanoparticle and biopolymer-based biomaterials for wound healing applications, Int. J Biol. Macromol. 115 (2018) 165-175. https://doi.org/10.1016/j.ijbiomac.2018.04.003

[75] M. Naseri-Nosar, Z.M. Ziora, Wound dressings from naturally-occurring polymers: A review on homopolysaccharide-based composites, Carbohydr. Polym. 189 (2018) 379-398. https://doi.org/10.1016/j.carbpol.2018.02.003

[76] T. Krupp, B.D. dos Santos, L.A. Gama, J.R. Silva, W.W. Arrais-Silva, N.C. de Souza, M.F. Américo, P.C.J.I.de Souza Souto, Natural rubber-propolis membrane improves wound healing in second-degree burning model, Int. J. Biol. Macromol. 131 (2019) 980-988. https://doi.org/10.1016/j.ijbiomac.2019.03.147

Advanced Applications of Polysaccharides and their Composites   Materials Research Forum LLC
Materials Research Foundations **73** (2020) 86-135   https://doi.org/10.21741/9781644900772-4

Chapter 4

Applications of Chitosan Composites in Pharmaceutical and Food Sectors

Rabinarayan Parhi

Department of Pharmaceutical Sciences, Susruta School of Medical and Paramedical Sciences, Assam University (A Central University), Silchar-788011, Assam, India

bhu_rabi@rediffmail.com

Abstract

Despite of many advantages including biocompatibility, biodegradability and nono-toxicity, chitosan alone suffered in terms of application in pharmaceutical and biomedical fields because of low mechanical strength and poor thermal stability. Therefore, more attention is being given to chitosan based composites to improve above mentioned properties and to modify its drug release potential; chitosan is being blended with other natural or synthetic polymers or by incorporating nano-fillers to either obtain simple composites or nanocomposites. More importantly, these composite materials based on chitosan can be moulded into various shapes and forms such as hydrogel, films, fibers, microspheres, nanospheres, scaffold, beads, sponges and solution to suit different pharmaceutical and biomedical applications including drug delivery systems, tissue engineering and wound healing, and food packaging.

Keywords

Chitosan, Composite, Biodegradable, Biocompatable, Microsphere, Scaffold

Contents

Abbreviations

DDS: Drug delivery systems

NPs: Nanoparticles

GIT: Gastro-intestinal tract

PLGA: Poly(d,l-lactide-co-glycolide)

TPP: Tripolyphosphate

GA: Glutaraldehyde

BSA: Bovine serum albumin

PVA: Poly (vinyl alcohol)

TB: Tuberculosis

TMC: N-trimethyl chitosan

PEG: Polyethylene glycol

TRI: Tripterine

PNIPAM: Poly(N-isopropylacrylamide)

CMC: Carboxymethyl chitosan

LCST: Lower critical solution temperature

PU: Polyuratheane

CP: Calcium phosphate

CPC: Calcium phosphate cement

PMMA: Polymethylmethacrylate

RGD: Arginylglycylaspartic acid

IPN: Interpenetrating polymer network

ELISA: Enzyme linked immunosorbent assay

PDLLA: Poly-d, l-lactic acid

NGF: Nerve growth factor

SCAs: Stem cells derived from adipose tissues

MI: Myocardial infraction

GO-Au: Graphene oxide gold nanosheets

SEM: Scanning electron microscopy

MTT: 3-(4,5-dimethylthiazol-2-yl)-2,5-diphenyl tetrazolium bromide

HepG2: Human heatoma cell line

PVC: Polyvinyl chloride

CNT: Carbon nanotubes

PVP: Poly (N-vinylpyrrolidone)

TiO_2: Titanium dioxide

MRI: Magnetic resonance imaging

HGC: Hydrophobically modified glycol chitosan

DOX: Doxurubicin

EPR: Enhanced permeability and retention effect

FITC: Fluorescein isothiocyanate

LSE: *Lepidium sativum* seedcake phenolic extract

FTIR: Fourier-transform infrared spectroscopy

FRAP: Ferric reducing ability power
DPPH: 2,2-diphenyl-1-picrylhydrazyl
PA: Protocatechuic acid

1. Introduction

A composition or composite material is one in which two or more constituent materials having different physical or chemical properties are combined to produce a material with properties/characterstics different from the individual components [1]. In the context of drug delivery, polymer composites are being widely used in recent times as these material produces properties substantially unattainable with copolymers and homopolymers. The properties in the focuss include higher mechanical strength to withstand pressure during use, control drug release, and temperature resistance. Presently, more attention is being given to biopolymer based composites because of desired properties such as biocompatibility, biodegradability and non-toxicity. Despite of advantages, biopolymers suffered in terms of application in pharmaceutical and biomedical fields due to their low mechanical strength and poor thermal stability [2]. To improve above mentioned properties along with to modify drug release, blending of more than one natural polymers or natural with synthetic polymer are performed or by incorporating nano-fillers in to either of the polymer (s) in order to obtain either simple composites or nanocomposites [3].

Chitosan is a well known natural polysaccharide. In recent times, chitosan has attracted many researchers not only from academia but also from industry, particularly in pharmaceutical field because it is safe, stable, biodegradable, biocompatible and sterilizable. Apart from above, chitosan also offers desirable pharmaceutical qualities such as ability to cross-link with different polymers, controlling the rate of drug release, bioadhesion, antibacterial properties, and activation of macrophages. However, chitosan is poorly available in nature and obtained from the cells of certain fungi. Whereas chitin, source of chitosan, abundantly present in the cell wall of algae and fungus, and cuticles of insects, crabs and shrimps [4]. Therefore, the main source to obtain chitosan is from the chitin by N-deacetylation process. Chitosan is composed of two components namely: N-acetyl D-glucosamine and 2-amino 2-deoxy-β-D-glucopyranose [5,6]. The glycosaminoglycans is a major component of extracellular matrix of bone and other tissues in the human body [7,8]. The functionality of chitosan is determined on the basis of positive amino groups and negative hydroxyl groups presence [7,9]. Modification of chitosan molecules can be carried out easily due to the presence of amino group at the C2 position of the monomer ring [10]. Despite of many advantages, chitosan alone has

Advanced Applications of Polysaccharides and their Composites   Materials Research Forum LLC
Materials Research Foundations **73** (2020) 86-135   https://doi.org/10.21741/9781644900772-4

limited use due to its weak mechanical properties and uncontrolled dissolution along with solubility issues [11,12].

Chitosan is one of such polymers which usually react with other polymers or small molecules to transform them into composites or derivatives. This modification under mild condition is possible due to the presence of free amino groups on the chitosan molecule. There are several approaches described to combine chitosan with natural/synthetic polymers to obtain desired properties including (i) electrostatic interactions, (ii) hydrogen bonding, (iii) hydrophobic association, and (iv) using crosslinker [13,14]. More importantly, these composite materials based on chitosan can be moulded into various shapes and forms such as hydrogel, films, fibers, microspheres, nanospheres, scaffold, beads, sponges and solution to suit different pharmaceutical and biomedical applications such as drug delivery systems (DDS), tissue engineering and wound healing (Fig.1) [15,16]. Therefore, this chapter discussed all the pharmaceutical and food packaging aspects of chitosan composites along with drug delivery qualities of chitosan.

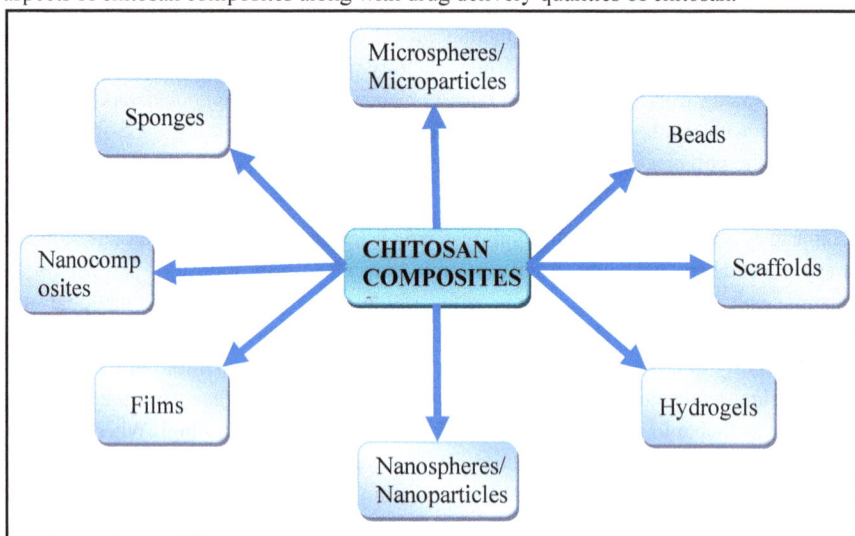

Figure 1. Chitosan based composites fabricated into various drug delivery forms.

2.   Applications of chitosan composites in pharmaceutical sectors

Chitosan composites are widely being used in various pharmaceutical and biomedical sectors including drug and gene delivery, tissue engineering, implants, contact lens, wound healing, bioimaging and cancer therapy, and in food sectors (Fig. 2).

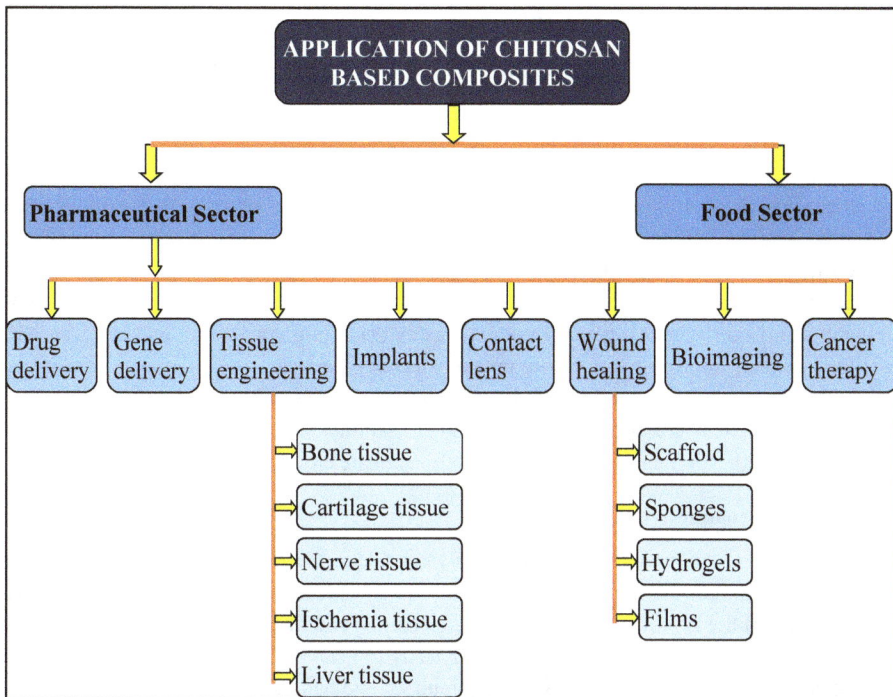

Figure 2. Schematic diagram showing various applications of chitosan composites in pharmaceutical and food sectors.

2.1   Application of chitosan composite in pharmaceutics/drug delivery

The basic concept of controlled drug delivery is to provide a precise release rates or spatial targeting of active ingredients to specific body site. Drug release period may vary from few hours to few months depending on DDS and their applications. In order to achieve this, the therapeutic agent is incorporated in a suitable polymeric network that can be single polymer or polymeric composite [17,18]. Chitosan in various forms, such as tablets, capsules, micelles, microspheres, hydrogels, films, wafers, nanoparticles (NPs) and composites are widely used to deliver drugs of diverse classes along with vaccines and nucleic acids. Because of higher viscosity, high molecular weight chitosan exhibits prolong drug release [19]. The addition of polymer or NPs to chitosan matrix to form composites can further sustain the drug release and thereby improves the therapeutic efficiency and patient compliance.

2.1.1 Properties of chitosan suitable for drug delivery

2.1.1.1 Anionic properties

An ionic interaction means of controlled drug delivery can be thought of when other mechanisms of drug release such as dissolution, diffusion or erosion are ineffective. However, an anionic drug can only be considered for this because of cationic characteristics of chitosan. Complex obtained from chitosan and anionic drug not only control the release of encapsulated drug but also make the complex more stable. Stable and significantly improved drug uptake was evident from enoxaparin and chitosan NPs [20]. Further improvement of drug uptake was not observed when composite of chitosan with other inorganic polymeric anions and multivalent anions such as sulfate or tripolyphosphate are used. However, higher stability was achieved for composite of chitosan with anionic polymer such as pectin, alginate etc. [21].

2.1.1.2 Mucoadhesive properties

Chitosan is a well known polycationic polymer due to the presence of -NH_2 group on its glucosamine unit. In swollen condition, this group interacts with sialic acid component (anionic) of the mucin chain and establishes electrostatic interaction. In addition, chitosan also develop other non-covalent bond such as hydrogen bond with mucus layer leading to its adherence to both soft tissues such as mucous and hard tissue such as epithelial tissue [22]. Furthermore, trimethylation of the primary amino group and immobilization of thiol groups of chitosan led to higher cationic character and mucoadhesive properties, respectively [23]. Enhancement of mucoadhesive properties of chitosan enable any DDS to adhere mucous layer, present all along the gastro-intestinal tract (GIT) and in other body cavities, for considerable period of time resulting in the penetration enhancement and subsequent absorption of active ingredient into systemic circulation [24,25].

2.1.1.3 Gelling properties

In hydrogel form, chitosan shows pH-dependent *in situ* gelling properties. *In situ* formulation is one that exist in the liquid/solution form in container and but converted to gel form on contact with body fluid (e.g. tear fluid). As a result of that contact time of the formulation increased and subsequently, drug absorption with an increased duration of therapeutic effect. The mechanism to trigger the transition of liquid phase to gel phase depends on the particular polymer employed. Mainly, change in temperature, pH, ion sensitivity or ionic strength are responsible. There was a formation of liquid state when chitosan is combined with polyacrylic acid at lower pH (6.8). The resulted liquid state underwent a fast transition to viscous gel state at pH 7.4 [26]. Further improvement in *in situ* gelling properties of chitosan can be carried out by modification such as thiolation. There was an increase in viscosity of 16,500 times using 1% chitosan-ethioglycolic acid conjugate over a period of 20 minutes [27].

2.1.1.4 Gene expression properties

Modified chitosan such as self-branched and thiolated chitosan are exhibited enhanced gene expression property. Self-branched chitosan doubled the gene expression compared to Lipofectamine and 5 times compared to Exgen [28]. The intra-chain disulfide bonds in thiolated chitosan enhanced the level of stability of complex between chitosan and plasmid NPs. In addition, plasmid was released in the target site due to the cleavage of disulfide bond inside the reducing condition of cytoplasma [29]. Furthermore, the PEGylated chitosan NPs, and chitosan and cyclodextrin complex are considered as promising tool in DNA-based drug delivery [30].

2.1.1.5 Permeation enhancing properties

The cationic property of the chitosan is responsible for the permeation enhancement of the incorporated drug. The mechanism is believed to be the interaction of positive charges on chitosan and negative charges on cell membrane leading to structural re-organization of tight junction related proteins [31]. It is evident from the 2-fold increase in gancyclovir oral bioavailability when co-administered with chitosan [32]. It was reported that the higher molecular mass and higher degree of deacetylation of chitosan resulted in higher epithelial permeability [33].

Indomethacin, a nonsterodal anti-inflammatory drug, was encapsulated in hydrogel microspheres formed from polyacrylamide grafted chitosan cross-linked with glutaraldehyde (GA). The polymer chain relaxation caused the initial drug release, whereas molecular diffusion due to full swollen polymer structure led to control release of drug [34]. Dysregulation of iron plays an important role in the development of various diseases such as bone, stoke, skin and muscle diseases, and neurological disorders. Deferoxamine is the first choice for the treatment of iron dysregulation. However, deferoxamine has limitation of short plasma half-life and poor absorption from gut wall. Therefore, to circumvent it deferoxamine was incorporated in to poly(d,l-lactide-co-glycolide) (PLGA) based microspheres and then loaded into chitosan and alginate based hydrogel. This composite hydrogel showed sustained drug release compared to microsphere alone, which attributed to the ability of hydrogel to control the drug release after the drug is being released from microspheres inside the composite gel [35].

Magnetic particles (Fe_3O_4) and chitosan fiber based composite NPs were developed using cross-linkers such as sodium tripolyphosphate (TPP) and GA in order to incorporate model protein drug bovine serum albumin (BSA). Thereafter, with poly(vinyl alcohol) (PVA) the resulted nanocomposites were electrospun. After 30 h of release study, there was 60% of BSA release indicating a controlled drug release profile. This is due to magnetic properties along with fibrous networks of chitosan and PVA [36].

Tuberculosis (TB) is a serious infectious disease which mainly attacks the lungs. Therefore, the main goal for the treatment of TB is to achieve enhanced drug concentration in the pulmonary tissue which will improve the therapeutic efficacy of the drug and minimize the systemic side effect [37,38]. Rifabutin, an anti-TB drug, was incorporated in composite microspheres formed of chitosan and ethyl cellulose with genipin as cross-linker. *In vivo* studies on rats did not show any significant difference in drug release between chitosan microspheres and the genipin cross-linked microspheres from 4 to 24 h. This was ascribed to the presence of lysozyme of airways that weakening the genipin cross-linked microspheres. Thereafter, from 24-192 h, drug release from the genipin crosslinked microspheres exhibited higher rate of drug release and after 24th day rifabutin concentration in lungs was 1.54±0.23 g mL⁻, which was higher than that of rifabutin release from microspheres (0.68 ±0.15 g mL⁻) without genipin cross-linking [39].

In one study, three different molecular weight grade of chitosan (low, medium and high) were used to synthesize N-trimethyl chitosan (TMC) chloride and then, they were used to develop nasal based hydrogel by co-formulating with glycerophosphate and polyethylene glycol (PEG). Out of them, medium and high molecular weight TMC based hydrogel demonstrated good water retention capacity and strong mucoadhesive potential with relatively short sol-gel transition temperature (32.5°C). In addition, the TMC with medium average molecular weight exhibited most promising results with favourable rheological and mucoadhesive properties and 7 minutes of sol-gel transition time [40].

An *in vivo* study on chitosan oligomer and zidovudine composites was investigated. The pharmacokinetic study demonstrated higher retention time in mice for the composites compared to drug alone and the composites were found to accumulate in brain, kidney, spleen and lung following their *in vivo* administration. Therefore, it was concluded that the composites can be developed into renal targeting DDS [41]. An innovative ethyl cellulose based microspheres incorporated chitosan film was developed with an intention to control the ciprofloxacin HCl release. In the first step, microspheres of ciprofloxacin HCl was developed using ethyl cellulose and then dispersed in chitosan solution followed by casting and solvent evaporation to obtained composite film. These films showed good compatibility between drug and film matrices due to hydrogen bonding and ionic interactions leading to controlled release of drug [42].

Tripterine (TRI) is a herbal drug extensively used in the treatment of cancer. However, its low bioavailability is the biggest challenge for the successful use in the clinical set up. To circumvent TRI pharmaceutical constraint, phytosomes were developed and then coated with protamine (PRT) layer using electrostatic assembly process. The resulted PRT-TRI phytosomes (size of 250 nm) were efficiently incorporated into composite sponges with chitosan and HPMC for transmucosal buccal drug delivery. PRT coated phytosomes based composite sponges exhibited sustained release profile with superior mucoadhesive characteristics compared to sponges contain PRT uncoated phytosomes. This was attributed to possible interaction between PRT and chitosan. There was a 2.3-fold increase in flux across chicken pouch mucosa and 244% higher relative bioavaiability was observed with PRT coated tripterin sponges compared to its counterpart [43].

In one investigation, multi-responsive smart composite microspheres loaded with indomethacin and magnetic NPs were developed using thermoresponsive polymer poly(N-isopropylacrylamide) (PNIPAM)) and carboxymethyl chitosan (CMC) via *in situ* free radical polymerization and emulsion cross-linking technique. Resulted composite microspheres demonstrated good response to applied magnetic field. The release of entrapped drug was increased once their solution temperature exceeded lower critical solution temperature (LCST) (50°C) and solution pH was in basic range (pH=11.0). Above features make the composite microspheres a potential controlled drug release vehicle [44]. In another study, same multi-responsive composite microspheres based on PNIPAM and chitosan was developed to deliver model drug berberine HCl in controlled manner. The encapsulation efficiency was found to be highest (73.5%) in acidic environment (pH = 4.0) compared to neutral (20.3%, pH = 6.9) and alkaline (15.1%, pH = 9.2) environment. Thermo-induced swelling and shrinking of the microspheres were observed above and below LCST of PNIPAM (32°C) indicating thermo-responsive properties of the composite microsphers [45].

Electrospun hybrid nanofibers of model drug danshensu was prepared in two steps: (i) chitosan coated illite particles loaded with danshensu were developed in the first step and (ii) then, these particles were dispersed into PLGA and polyuratheane (PU) solution to finally fabricate nanofibres. These hybrid nanofibers were found to be uniform and smooth morphology and demonstrated good biocompatibility. Particularly, these composite nanofibers considerably decreased burst-release and showed sustained release profile of the incorporated drug [46].

Chitosan coated mesoporous silicon microparticles were developed for controlled delivery of peptide and protein drugs. Mesoporpous silicon particles showed high affinity to BSA and moderated affinity to insulin. More than 80% of the insulin was released from the microparticles in 45 minutes, but without burst release (<20%). Coating of microparticles with chitosan had provided two advantages; (i) sustained release of drug for considerable period of time and (ii) enhancement of other desired features such as mucoadhesion and permeation [47].

2.2 Chitosan composite in gene delivery

Compared to polygalactosamine-DNA complexes, chitosan-DNA complexes are more effective in non-viral gene therapy and more importantly very easy to synthesize. However, their use is curtailed by lower transfection efficiency and stability, and the factors responsible are pH of the culture media, molecular weight of chitosan etc. [48-50]. One approach to improve the efficiency is to develop polymer complexes, called as polyplexes, through electrostatic interaction between negatively charged nucleic acid and a polycation. Due to the formation of polyplex, primary amine group on the glucosamine repeating unit control the charge density. This in turn depends on the pH as well as degree of acetylation. This resulted in the successful application of chitosan in non-viral gene delivery such as compensating defective genes by gene silencing and production of beneficial proteins or DNA vaccines [51]. In another way, quaternization of chitosan such as TMC improved the stability of ionic complexes with DNA [52]. In addition, the combination of quaternization and attaching thiol group on TMC produces mucoadhesive properties of TMC by disulfide bond formation with mucin proteins of the cell membrane. A novel injectable system composed of chitosan and pluronic composite hydrogel for gene therapy was successfully developed. Transfection studies on HEK293 cells demonstrated the release amount from the composite hydrogel showed higher transfection efficiency compared to pluronic based hydrogel [53].

2.3 Chitosan composite in tissue engineering

Tissue engineering is based on the concept of combining cells or tissues, engineered bioactive molecules/biomaterial and biochemical factors to replace, repair, or improve biological functions of tissues or organs that are injured beyond identification [54-57]. This involves various applications including repair, replacement of whole or part of tissues such as cartilage, blood vessels, bladder, bone, skin, muscles and nerve tissues [9]. Compared to old techniques such autografts and allografts, tissue engineering has many advantages including better availability, no pain at graft site, without better survival rate and lower cost [58,59]. Biopolymer, particularly chitosan and/or chitosan based biomaterials, have became extremely popular because of their nontoxic and biocompatibility properties. In addition, these materials showed certain structural and mechanical properties similar to that of damaged tissues [57]. Therefore, derivatives of chitosan such as composite or scaffolds are utilized in cell seeding and employed in the development of tissue analogous of different organs *viz.* bone, cartilage, nerve, bladder, liver and skin tissues [60]. Furthermore, chitosan based biomaterials can be developed into various forms such as gels, powders, films and scaffolds, which enhances their wide usability [56].

2.3.1 Bone tissue engineering

Bone is an important part of the human body, which is made up of basically two matrices: inorganic matrix made up of carbonated hydroxyapatite and organic matrices composed of collagen and non-collagenous proteins [61,62]. Bone tissue engineering involves the use of matrix with or without cells or biomaterials or a mixture of all of them to address bone defects or the regeneration of bones. Osteoconduction is an ideal property of chitosan in addition to biodegradability and biocompatibility for which it is widely employed in bone tissue engineering. Biocompatibility property of chitosan minimizes local inflammation; whereas osteoconduction provides space for the growth of the new bone tissues, facilitates neovascularization and allows nutrient transfer to the bone tissues [63,64]. Moreover, chitosan found to induce cell growth and enhance process of deposition of mineral rich matrix by osteoblast cells [60]. However, drawbacks of chitosan such as limited mechanical strength and stability issues do not allow it for use in its native form. To surmount above limitations, chitosan blended with other polymers have been extensively employed in tissue engineering. The proper blending of polymers with chitosan not only improve their mechanical property but also enhance cell growth and osteoconduction [65,66].

Numbers of studies were carried out to develop composites of calcium phosphate (CP) with chitosan to improve its properties. In one such study, CP was embedded in porous chitosan sponges to form 3D macroporous composite. It was observed that the nested chitosan sponge enhanced the mechanical strength of the composite by matrix reinforcement and able to preserve the osteoblast phenotype [67]. SEM study showed that pore size (nearly 100 μm) of chitosan sponges formed inside macroporous structure of the composite was favourable for the bone tissue in-growth. Furthermore, the composite exhibited good cell biocompatibility as osteoblast cells attached to composite proliferated on its surface and migrated into the composite. Matrix reinforcement was improved by incorporating beta-tricalcium phosphate and CP invert glasses to gentamycin-conjugated macroporous chitosan scaffold. It was found that the initial burst release of gentamycin sulfate was decreased to the reinforcement and sustained release was observed for greater than 3 weeks. In addition, osteoblast cells were grown and migrated into the scaffold which indicated better cellular compatibility [68]. In another investigation, macroporous scaffold having interconnected porosity of 100 mm was developed with the incorporation of hydroxyapatite or CP glass [69]. These composites of chitosan with CP demonstrated better clinical application in the future.

One promising material, calcium phosphate cement (CPC), is being used widely for craniofacial and orthopeadic repairs as it hardens *in situ* to convert into solid hydroxyapatite of complex cavity shapes without machining and has excellent osteoconductivity and reabsorbable property. However, lack of porosity and low strength of CPC curtails its wide use. Incorporation of chitosan into CPC increased its tensile strength from 3.3 MPa to 11.9 MPa and flexural strength by 39%. The work-of-fracture i.e. toughness of composite was found to be increased by two times compared to only CPC. Thus, composite may be employed in the repair of stress-bearing orthopedic and craniofacial repairs [70]. In one study, a novel biodegradable biomimetic composite with hydroxyapatite, chitosan and gelatin was developed by phase separation method. It was observed that highly porous scaffold (porosity of 90.6%) allowed better attachment of osteoblasts and proliferation on its surface. In addition, the composite showed good biomineralization after 3 weeks of study in culture medium [71].

A novel bioactive bone cement composed of hydroxyapatite, chitosan and polymethylmethacrylate (PMMA) was synthesized. Compared to PMMA based bone cement, the composite bone cement was more intrusive, biocompatible and osteoconductive, and has higher porosity. Furthermore, the developed composite bone cement did not show any cytotoxic characteristics [72]. CPC based scaffold with chitosan (15%) and polyglactin (20%) fibers were developed for the delivery of harvested human umbilical cord mesenchymal stem cells for bone tissue engineering. The scaffold showed higher flexural strength of 26 MPa compared to 10 MPa for CPC alone, which improved the resistance against fatigue and fracture. In addition, the scaffold exhibited better cell proliferation and viability [73].

In bone tissue engineering, nano-hydroxyapatite is being widely used because of its structural resemblance with natural bone. Mechanical properties of chitosan-based preparations are improved with the incorporation of nano-hydroxyapatite properties [74]. In one study, a scaffold composed of freeze gelated chitosan and hydroxyapatite was prepared and investigated for its suitability in bone tissue engineering. The scaffold of chitosan incorporated with hydroxyapatite exhibited higher stability compared to scaffold of chitosan alone [75]. In another study, a novel composite scaffold composed of chitosan-gelatin with nanohydroxyapatite-montomorillonite were developed employing freeze-drying method. Resulted composite demonstrated a decreased in swelling, porosity and degradation but enhanced the mechanical properties. The decrease in swelling was due to the nano-hydroxyapatite and montmorillonite based temporary barrier formation, which prevented the easy entry of water molecules into the composite structure. In addition, these nanomaterials bind with –COOH or –NH$_2$ groups of gelatin molecule which resulted in the reduction of its hydrophilicity. Furthermore, the nano-sheets of montomorillonite reduce the interaction between water and polymers. The porosity of the composite was decreased due to incorporation of nano-materials into it. In addition, they reduce the degradation rate of the composite simultaneous with the enhancement of mechanical property [76]. The incorporation of hydroxyapatite filler having titania powder into chitosan-gelatin copolymer mixture resulted in the formation of a biocomposite of hydroxyapatite-titania/chitosan-gelatin. The composite containing 10% of titania exhibited better resemblance with natural bone compared to composite containing 30% titania, whereas higher mechanical properties was observed for composite having 30% of titania [77].

Modified chitosan, N-(2-carboxybenzyl)chitosan, was used along with various weight percentage of titania NPs and bioglass 45S5 to form 3D macro-porous hybrid scaffolds by freeze drying technique. The modification of chitosan and the addition of inorganic fillers contributed to the increase in mechanical strength such as comprehensive strength and modulus five times higher than chitosan alone. Further increase (12-fold) in mechanical strength was observed by the addition of 2.5 wt% of bioglass to scaffold containing 2.5wt% of titania. All the developed scaffolds were found to be bioactive, safe, and contain interconnected pores (150-300µm). Furthermore, sponge like behavior was observed for all the scaffolds. In addition, the study on mesenchymal cells and human embryonic kidney cell line demonstrated that these cells attach and migrate into the scaffolds well [78]. In another investigation, biodegradable nano-composite scaffolds composed of chitosan and gelatin along with bioactive glass ceramic NPs was developed for alveolar bone regeneration applications. Investigation on the scaffold revealed that the addition of NPs to the scaffold decreased their degradation and swelling behavior, while their protein adsorption capacity was increased. In addition, nanocomposite scaffold was deposited with higher amount of mineral and the rate of deposition further increased with the increase in time of incubation [79].

A nanocomposite based on chitosan and carbon nanotubes (f-MWCNT) were developed by freeze-drying technique. It was observed that nanocomposite uniformly dispersed in chitosan matrix. Cell proliferation and cytotoxic effects of resulted nanocomposites were studied on human osteosarcoma cell line (MG-63) employing MTT assay method. The result showed that cell proliferation and mineralization of cells cultured on nanocomposite were higher compared to chitosan scaffold alone [80]. Composites based on chitosan and β-tricalcium phosphate were developed and the resulted composite found to show histocompatibility with mesenchymal stem cells of Beagle. In addition, composites demonstrated higher potency in augmenting process of osteogenesis and restoration of bone defects [81]. A novel scaffold with porous structure was developed with hydroxyethyl chitosan and cellulose. The scaffold showed promising result on the basis of attachment, spreading and proliferation of osteoblastic cells . Therefore, the resulted composite was considered as a matrix for bone tissue engineering and was also tried for cartilage engineering [82].

2.3.2   Chitosan in cartilage tissue engineering

The integration of cells, biomaterials and factors with extremely particular functions and attributes are the basis of cartilage tissue engineering. It follows steps of cell/tissue culture: first step involves number of cells increase and the next step involves the induction of cells to form specific cartilaginous phenotypes. There were number of biomaterials used successfully for this purpose, including polysaccharides *viz.* chitosan, alginate, and hyaluronan, proteins (e.g. elastin, collagen, keratin and fibroin), and polysters (e.g. poly(hydroxybutyrate) [83-89]. In the selection of biomaterials, the important aspects should be kept in mind is its biodegradation. This implies that the material should have the potential to be degraded by enzymes so that new tissue grow in that place and the degradation products should not produce any inflammatory response or immunoresponse within the newly formed tissue. Among various biomaterials, chitosan is preferred as composite or scaffold material because of its structural similarity with various glycosaminoglycans present in the articular cartilage [90,91]. There are many factors to be considered in the designing of composite or scaffold for cartilage tissue engineering including their pore structure (size, shape and distribution), elasticity, surface energy, mobility characteristics, chemical structure, sensitivity towards environmental factors such as pH, temperature and stress, biocompatibility and biodegradation [92,93].

A homogenous composite hydrogels based on chitosan and carrageenan was developed using cross-linking agent epichlorohydrin. Prepared composite hydrogel exhibited higher viability, adhesion, proliferation and differentiation of ATDC5 cells *in vitro*. Increase in chitosan content in the hydrogel significantly increased the chondrogenic differentiation of ATDC5 cells. Thus, the composite hydrogel based on chitosan could be used as a promising cell carrier in cartilage repair [94]. Three polymers such as chitosan, alginate and hyaluronan based composite material with or without covalent bonding with arginylglycylaspartic acir (RGD) associated protein in order to enhance cellular adhesion of chitosan. In *in vitro* study it was observed that the cell-seeded composite demonstrated neocartilage formation. Whereas a partial repair cartilage defects was witnessed after one month both in the absence and presence of RGD. These results indicated the prepared composite has the potential for not only cartilage repair, but also cartilage generation [95].

Silk matrix reinforced with chitosan microparticles at different ratio generate composites having better visco-elastic matrix property which not only improved redifferentiation of caprine chondrocytes, but also retain higher glucosaminoglycan. The latter improved the aggregate modulus and has ability to be employed in cartilage tissue engineering [96]. In another investigation, porous microcarriers of CMC was first prepared and then coated with collagen nanofibers to obtain an injectable composite for cartilage tissue engineering. The *in vitro* cell culture study demonstrated that chondrocytes not only adhered, but also proliferated and differentiated on the composite microcarriers. It was further observed that within three days post-seeding chondrocytes grew to confluence on the microcarriers and after 7 days several confluent of macrocarriers found to attach with each other leading to tissue like aggregates [88].

A promising injectable composite hydrogels including methacrylated glycol chitosan and hyaluronic acid was developed by employing riboflavin and visible light as photo-cross linking agents. By increasing the irradiation time from 40 to 600 s, not only reduced the encapsulation of chondrocytes but also improved the compressive modulus of the hydrogel from 11 to 17 kPa. It was also observed that the incorporation of hyaluronic acid to chitosan improved the deposition and proliferation of cartilaginous extracellular matrix by the incorporated chondrocytes [84]. Chitosan and alginate based novel semi-interpenetrating polymer network (IPN) scaffold was developed by freeze-drying technique. The resulted IPN scaffold demonstrated attachment and proliferation of ATDC5 murine chondrogenic cells. Among various combinations of alginate and chitosan, 50:50% (v/v) ratio exhibited promising result in terms of structural analysis and cell-based functional screening in cartilage tissue engineering *in vitro* [97].

2.3.3 Chitosan in nerve tissue engineering

Mature neurons do not undergo cell division. Therefore, any nerve injuries can lead to malfunctioning of nervous system in other parts of the body and it also complicate the situation as healing and rehabilitation of mature neurons are difficult without proper material [98,99]. Thus, direct regeneration of nerve fibers into proper endoneurial tubes is the best way out. For this regeneration, there are two types of strategies employed: (i) employing grafting and tubulization method for bridging and (ii) end-to-end nerve stumps suturing. Between two, tubulization technique has been considered as more efficacious, as it circumvent the stress across the neurons repair site [100]. However, the internal surface area of an artificial tube will not be enough for the coherence between nerve fibers and Schwann cells [9]. Therefore, there is a need for biodegradable matrix which can offer a cellular and molecular skeleton for neurite and Schwann cells migration through the nerve gap. In this context, chitosan and its composites are suitable for the regeneration of nerve because of their biocompatible and biodegradable characteristics [101]. It has been proved that chitin and chitosan are potential materials for improved cell adhesion and neurite outgrowth [98].

In one study, attachment, differentiation and growth of nerve cells were enhanced on the composite based on chitosan and poly(l-lysine) compared to chitosan membrane alone. The affinity of nerve cell towards composite material was due to the enhanced hydrophilicity caused by the positive charges and hydroxyl groups on the chitosan [102].

A series of composite films were developed by mixing chitosan with three different polycations such as poly(L-lysine), polyethyleneimine and poly-L-ornithine and studied the impact of surface topography on the adhesion, proliferation and differentiation of nerve cells on it. PC12 cell culture was used to study the above effect. It was observed that the cells demonstrated higher level of adhesion, proliferation and differentiation on all types of developed composite films compared to granules, particles or island surfaces [103]. Composite films of chitosan and gelatin were developed and observed that all the composite films exhibited higher percentage of elongation at break and lower Young's modulus compared to chitosan film. The enzyme linked immunosorbent assay (ELISA) test demonstrated higher amount of fibronectin adsorption on the composite films than that of chitosan film. In addition, the affinity of nerve cell towards the composite materials was higher *in vitro*, when PC12 cell culture was used. The cell culture also showed that the composite film composed of 60 wt% of gelatin demonstrated higher cell differentiation and extended longer neuritis than the chitosan film. The result was attributed to the soft and elastic complex nature of the composite material [104].

An aligned nanoscale fibrous scaffold was developed by mixing poly(3-hydroxybutyrate) and chitosan (at 15 and 20%) in trifluoroacetic acid and then electrospinning the resulted solution. The average diameters of the aligned scaffolds were found to be 740.3 nm and 870.74 nm for 15 and 20% of chitosan, respectively. The hydrophilicity of scaffolds were found to be increased with the increase in chitosan concentration, and Young's modulus and tensile strength of aligned form scaffold were increased compared to formulations having random nanofibers. Out of two combinations, the formulations containing 15% of chitosan exhibited higher suitability in terms of nerve tissue regeneration [105].

Nerve conduits formed of poly-d,l-lactic acid (PDLLA), chondroitin sulfate and chitosan were developed to study its potential to repair damaged nerve and the effect of nerve growth factor (NGF). The NGF was immobilized on to the conduits with carbodiimide and then conduit was implanted in the created defect in the sciatic nerve of rats. After three months of implantation, the nerve conduits demonstrated a promising recovery in the disrupted nerves and it also prevents the ingrowth of connective tissue into the conduits due to its compactness. Furthermore, the conduit has high mechanical strength and biodegradable products were non-toxic. Thus, the prepared conduit can be used for nerve repair in peripheral nerve defects [106].

2.3.4 Chitosan in ischemic tissue engineering

Stem cells derived from adipose tissues (SCAs) were widely used in the area of regenerative medicine research not only because of its easy accessibility, but also due to its ability for multi-lineage differentiation and self-renewal [107,108]. In addition, it acts as angiogenic growth factors including hepatocyte growth factor and vascular endothelial growth factor [109]. However, direct injection of SCAs in to myocardium for the purpose of ischemic tissue repair and regeneration led to the cell death within 72 h of transplantation. Therefore, the paramount importance is to select a suitable biomaterial to maintain its angiogenic property along with its viability. Out of many natural polymers, chitosan was found to be stand out polymer as it is capable of enhancing the SCAs cell survival. However it lacks the required mechanical properties. This led to the development of chitosan composite with other polymers where both mechanical property as well as SCAs cell survival can be addressed [110,111].

Common symptoms after myocardial infraction (MI) are improper electrical impulse transmission and abnormal conduction. This resulted in the disruption of electrical communication between adjacent cardiomyocytes. In order to improve the communication, carbon nanofibers were incorporated into porous chitosan scaffold. The resulted composite bring about improved manifestation of cardiac-specific genes in neonatal rat heart cells which are responsible for muscle contraction and electrical coupling. The increase in cardiac tissue contraction due to incorporation of carbon nanofibers in chitosan scaffolds was attributed to enhanced transmission of electrical signals between the cardiac cells [112]. In another investigation, a biodegradable scaffold of chitosan incorporated with graphene oxide gold nanosheets (GO-Au) was developed. Addition of GO-Au resulted in increase in electrical conductivity by two fold and also supported better cell attachment and growth without cell toxicity. In MI induced rat, above composite found to increase cardiac contractility and restored ventricular activity [113]. More recently, proposed chitosan and gelatin composite based thermosensitive hydrogels demonstrated promising results to accelerate ischemic tissue regeneration by incorporating adipose-derived stem cells within it [109].

2.3.5 Chitosan in liver tissue engineering

Presently, liver transplantation seems to be the best option to treat various liver disorders. However, there are very few people that wish to donate liver for transplantation. Therefore, it is necessary to develop bio-artificial liver which is only possible *via* tissue engineering. First step in this process is to get suitable biomaterial. In this context, chitosan can be used as a scaffold for hepatocyte culture because of its structural semblance to glycosaminoglycans [114]. To improve the mechanical property of chitosan, composite materials with other polymers or nanomaterials could give desired properties.

In one such study, polymeric composites of chitosan and collagen were developed for the regeneration of liver. These composites showed desired mechanical strength. Moreover, hepatocyte culture and platelet deposition experiments demonstrated excellent cell and blood compatibility. Thus, it can be concluded that above composites are promising biomaterials for the implantable bio-artificial livers [115].

To continue with higher degree of functions pertaining to liver, a perfect extracellular matrix for hepatocytes culture were developed with scaffold composed of oxidized alginate cross-linked with galactosylated chitosan. Scanning electron microscopy (SEM) analysis revealed that the resulted composites have extremely pervious surfaces (average pore size of 50-150µm) and internal average pore size of 100-250 µm. The porosity of about 70% was observed for the developed material. Further, the equilibrium swelling and *in vitro* degradation rate of the scaffolds were decreased with the increase in oxidized alginate content, whereas thermal stability slightly increased. The hepatocyte culture on this material showed that the cells have taken typical spheroidal morphology and presented perfect integration with multicellular aggregates [116].

A series of experiments were performed to find out the best co-culture condition of hepatocytes in developed highly porous (150-200 mm pore size in diameter) hydrogel scaffold composed of galactosylated chitosan and alginate as artificial extracellular matrix. The resulted galactosylated chitosan/alginate and chitosan/alginate films demonstrated higher cell adhesion of 72.7 and 45 %, respectively, compared to only alginate film (28.5%). After 10 days of incubation, cell viabilities on galactosylated chitosan/alginate sponge were found to be 81.3%, which was higher than that of alginate film (72.7%). Hepatocytes were found to be aggregated as multicellular spheroids on the scaffold sponges with diameter enlarged up to 100 mm in 36 hr and these hepatocytes expressed connexin 32 and E-catherin genes connected to cell-cell adhesion. Thus, it was concluded that the hepatocytes spheroids which were developed on the scaffold could enhance liver specific functions and be helpful in developing a bioartificial liver [117].

A three-dimentional water stable scaffold composed of chitosan and silk fibroin were developed by freeze-drying technique. The resulted scaffold was observed to be water stable i.e; swelling to only limited extent based on the composition and demonstrated homogenous porous structure ranging from 100 to 150 µm with porosity above 95%, when the combined polymer percentage is below 6 wt%. It was also proved from the 3-(4,5-dimethylthiazol-2-yl)-2,5-diphenyl tetrazolium bromide (MTT) assay that the scaffolds can promote proliferation of human heatoma cell line (HepG2) significantly [118]. In another study, a series of composites with different ratio of chitosan and silk fibroin (4:6, 8:6 and 16:6) were developed to support hepatocyte attachment. Out of three combinations, the scaffold containing 4:6 ratio of chitosan to silk fibroin exhibited better results in terms of porosity, water absorption expansion rate and consistent aperature size (50-50 µm). In addition, the same scaffold showed significantly higher number of living cells compared to other two compositions [119].

2.4   Chitosan composite in preparation of implants

The events such as calcification, thrombosis and bacterial growth can pose a real problem for the functioning of implant leading to decrease in its life time and finally implant explanation. Thrombosis occurred due to activation of coagulation cascade followed by the adsorption of calcium, plasma protein, and platelets in less contact time between implant and blood. Whereas, in case of longer contact time, calcification of implants placed in breast, valves of heart and as stents, may lead to the deposition of various calcium salts. Growth of bacteria such as *Porphyromonas gingivalis* and *Streptococcus mutans* on the implants resulted in the accumulation of plaque [120]. One of the successful ways to deal with above problems is to coat clinical implants with biocompatible materials which would act as a biointerface. Chitosan among many polymers seems to be a gifted substance for above coating purpose because of its bicocomptibility and biodegradable characteristics [121]

In one such effort, coating material composed of Ag conjugated chitosan NPs was developed for titanium dental implants. After extracting from natural source (*A. flavus Af09*), chitosan was conjugated with Ag NPs. These coating NPs inhibited the adhesion and growth of two main microorganisms of teeth such as *S. mutans* and *P. gingivalis*, thereby decrease the biofilm formation. In addition, the material did not cause any cytotoxicity as extracted from natural resources [122]. In another effort, implatable tablets of risedronate sodium were developed with chitosan and polyvinyl chloride (PVC) based matrices. Compared to PVC based tablet, chitosan based tablet demonstrated poor compressibility leading to rapid disintegration of tablets. However, chitosan based tablets showed continuous drug release up to one week, which may be due to strong drug and carrier interaction [123].

Nanocomposite implants based on chitosan and carbon nanotubes (CNT) incorporated with hydroxyapatite to deliver calcium ions for peripheral nervous tissue regeneration. Incorporation of either single walled or multiwalled CNT enhanced mechanical characteristics of the implants. The developed implant was biocompatible as proved by *in vitro* cytotoxic studies on various cell lines. Therefore, the present implants could serve the patients suffering from peripheral nerve damages [124].

Composite implants of ciprofloxacin in the form of pellets were developed with chitosan and bovine hydroxyapatite (30:70) as matrix forming materials and GA as cross-linking agents. Addition of GA decreased the material crystallinity, thereby decreased mechanical strength of implants. However, GA inhibited the burst release of the drug and controlled the drug release profile. This was attributed to the decrease in porosity, water absorption capacity and swelling ratio of the implants. Thus, the present implants could be used to treat osteomyelitis for 30-days [125].

2.5   Chitosan composite in preparation of contact lenses

The anatomical and physiological restrictions of the eye have made drug delivery a challenge into eye tissues as these restrictions does not allow correct therapeutic concentration of drug to reach the desired site of action. Therefore, there are many dosage forms including liquid (eg. eye drops), semisolids (eg. gel, ointment) and solids (eg. inserts) have been recommended by clinical practitioners. In addition, vesicular systems including liposomes, niosomes etc. and polymeric and lipid NPs are also being widely used to treat eye ailments. Among all, therapeutic contact lenses seems to be very attractive DDS as these are designed such that it will cling to the eye surface due to surface tension without creating much discomfort to the patient [126]. However, these drug delivery devices suffer due to burst release at initial stage and thereafter drug release at subtherapeutic level. To tackle above problem and release the incorporated drug in a controlled manner, polymeric composite films of natural polymers including chitosan can be a better option as drug delivery contact lens device [124].

In one study, series of composite films were developed using chitosan and gelatin by solvent evaporation technique. The resulted contact lens demonstrated higher oxygen permeability, transparent, flexible and biocompatible. Moreover, the gelatin in the lenses improved the water absorption and enhances solute and oxygen permeability across the lens [127].

2.6   Chitosan composite in wound healing

The procedure of wound healing in living organism involves the combined action of cell and matrix components. It is the combination of wound regeneration and restoration of tissue integrity. It has three overlapping phases such as inflammatory, proliferative and remodeling phases [128]. However, the healing stops after the inflammatory phase. It does not proceed normally that leads to accumulation of inflammatory mediators, macrophages and neutrophils. This lack of healing can result into complex secondary microbial infections. Therefore, initial step to forment wound healing is to cover the wound with appropriate innocuous dressing material having semi-permeable nature in order to safeguard the wound from exterior mechanical and microbial burden. In addition, it maintains moist environment in the wound area to stimulate initiation of healing process [129].

Among various biopolymers, chitosan and its derivatives continued to be first choice in wound healing treatment. This is due to their structural resemblance to essential wound rebuilding molecule glycosaminoglycans present as a constituent in the extracellular matrix. It promotes the activity of macrophages, fibroblasts, polymorphonuclear leukocytes and antibacterial activity [130,131]. In addition, they swell through liquid uptake resulting into a sticky matrix that seals the wound and controls bleeding (Fig. 3). Furthermore, the biodegradation product N-acetyl-β-D-glucosamine leading to the stimulation of fibroblast proliferation, deposition of collagen along with hyaluronic acid synthesis at the wound area [132]. The positive charge on the chitosan attracts the negatively charged red blood cell which resulted in the rapid clot formation above the wound.

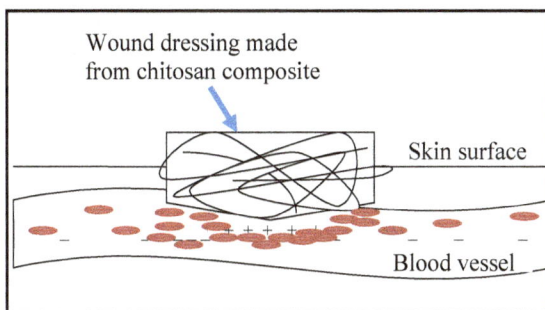

Figure 3. A schematic diagram representing the functioning of chitosan composite based wound dressing. Chitosan composite dressing works in two ways: (i) it absorbs fluid from blood after being placed on the wound site, which led to the sealing of wound and controls the bleeding and (ii) Negatively charged red blood cells are attracted by positively charged chitosan leading to formation of clot over the wound.

There were numerous studies reporting the use of chitosan based nanocomposite in the treatment of wound healing in various forms including scaffolds, sponges, hydrogels, films/membranes, and NPs.

2.6.1 Scaffold

A chitin hydrogel scaffold containing well dispersed Ag NPs was prepared to heal the wound. The resulted composite scaffolds exhibited antibactericidal activity against *Staphylococcus aureus* (*S. aureus*) and *Escherichia coli* (*E. coli*). With good blood clotting potential, these composite scaffolds demonstrated good cell surface adhesion when epithelial cells (Vero cells) were used [133]. In another study, scaffolds composed of chitin and Ag NPs was developed for wound healing applications. These scaffolds exhibited superb antibacterial potential against *S. aureus* and *E. coli*. Additionally, scaffolds demonstrated better blood clotting characteristics [134].

2.6.2 Sponges

Sponges are nothing but foams with open porosity. It is defined as dispersion of gas (basically air) into a solid structure/matrix to obtain a solid porous with soft and flexible properties [135,136]. The latter properties along with higher adhesive time, higher capacity to soak up wound exudates (more than 20 times of their dry weight) made them one of the better choices for the wound healing applications [137].

Promising composite sponges composed of cross-linked succinyl pullulan-CMC were successfully developed for wound healing applications. Resulted sponges exhibited desired water vapour transmission rate and absorb much liquid exudates due to high porosity. In addition, these composites showed no sign of cytotoxicity or hemolytic potential. Furthermore, histological examination indicated that above sponges hastened proliferation of fibroblast along with epithelial migration. *In vivo* studies on the back of ICR mice exhibited the effectiveness of sponges in healing full layer wound of skin defects [138].

2.6.3 Hydrogels

In one study, a flexible and microporous composite hydrogel composed of chitosan and nano zinc oxide was developed. Usually, chitosan does not exhibit its highest antibacterial potential. However, its activity can be augmented by incorporating an antibacterial agent in the form of nano zinc oxide. The resulted hydrogel exhibited higher degree of antibacterial activity and healing potential without causing cell toxicity. *In vitro* cytocompatibility studies demonstrated higher degree of cell viability and infiltration [139]. In another study, Ag NPs as effective antibacterial agent was incorporated into

Advanced Applications of Polysaccharides and their Composites   Materials Research Forum LLC
Materials Research Foundations **73** (2020) 86-135   https://doi.org/10.21741/9781644900772-4

gelatin and CMC hydrogels. Above hydrogel was prepared by radiation induced reduction followed by cross-linking at ambient temperature. Resulted hydrogel proved its antibacterial potential against *E. coli* and hydrogel containing 10 mM NP-Ag prevented greater than 99% of its growth. This result implied that the present hydrogel may be used for anti-inflammation wound dressing [140].

Natural product such as curcumin, obtained from the rhizomes of *Curcuma longa*, has shown its potential in enhancing wound repair and healing in diabetic patient. When nano-curcumin was incorporated in CMC and oxidized alginate hydrogel, it released from the gel in controlled manner and stimulated the simultaneous fibroblast proliferation, capillary formation and collagen production leading to rapid healing of wound [141]. In another study, a novel injectable hydrogel composed of chitosan and alginate loaded with nano-curcumin was developed. The release of nano-curcumin from the resulted composite gel significantly enhanced the process of re-epithelialization of epidermis and collagen deposition in the tissues having wounds. The resulted effect had led to almost complete wound closure after 14 days of treatment [142].

Based on the types of wounds and application of wound dressings, various kinds of anti-bacterial agents can be incorporated into hydrogel used for wound healing. Hydrogel of minocycline was developed with chitosan and PVA by freeze-thaw method. Among various compositions, the hydrogel containing 0.75% of chitosan, 5% of PVA and 0.25% of minocycline demonstrated suitable elastic, flexible and swellable qualities compared to PVA gel alone. This was due to comparatively weak cross-linking interaction between PVA and chitosan. Furthermore, histological investigation in wound induced rats showed huge collagen proliferations, microvessels and reduction in inflammatory cells, which resulted in efficient wound healing [143]. Another anti-bacterial agent, tigecycline, loaded (in the form of NPs) into hydrogel composed of chitosan and platelet rich plasma. The resulted hydrogel can effectively deliver the loaded tigecycline and demonstrated significant antibacterial activity against *S. aureus* [144].

2.6.4 Films

Composite films of chitosan are also being used for wound healing purpose. These are considered as medicated adhesive system which is capable of delivering required concentration of incorporated drugs to the desired site [145,146]. To be successful as drug delivery device, an ideal film should possess two important attributes namely adhesive and mechanical properties. A composite film of ciprofloxacin was prepared with thiolated chitosan and PNIPAM for wound dressing purpose. It was observed that thiol group on the chitosan enhanced its adhesion property and the mechanical property was found to be suitable for wound and burn dressing. The resulted film also exhibited cytocompatibility and released the drug in controlled rate for greater than 48 hr, which proved its anti-bacterial potential [147]. Composite films as dressing materials were developed with chitosan, poly (N-vinylpyrrolidone) (PVP) and titanium dioxide (TiO$_2$) and tested for their wound healing capability. The resulted composite film showed excellent antimicrobial efficacy against many microorganisms and biocompatibility against fibroblast cells (NIH3T3 and L929) *in vitro*. It further indicated accelerated healing of open excision type of wounds in albino rats, compared to soframycin skin ointment, conventional gauze and chitosan treated groups. Furthermore, the addition of TiO$_2$ based NPs enhanced mechanical properties of the composite film [148].

2.7   Application of chitosan composite in bioimaging

Drugs induced biochemical phenomena in cells are visualized and investigated employing important diagnostic tool such as bioimaging [56]. Biocompatible and biodegradable attributes of chitosan and its composites enable their application in various biomedical applications. In addition, chitosan composites as a new class of biomaterials which has potential for use in biomedical imaging with advanced applications due to their better physiochemical, mechanical and functional properties [149]. Bioimaging application is one of them where imaging agents such as Fe$_3$O$_4$, fluorescent materials were incorporated into chitosan or chitosan composites.

In one such investigation Fe$_3$O$_4$ as imaging material encapsulated into novel self-assembled NPs consisting of chitosan and linoleic acid conjugates for magnetic resonance imaging (MRI). There was a tremendous increase in *in vivo* molecular imaging of Fe$_3$O$_4$ due to its encapsulation in NPs. These composite NPs were used for hepatocyte targeted imaging [150]. In another study, CMC coated superparamagnetic iron oxide was developed for the visualization of human mesenchymal stem cells employing MRI. The solubility of chitosan in aqueous medium was enhanced by carboxymethylation, which led to better dispersion of coated particles. This resulted in higher labeling efficiency with identification of low numbers of labeled cells. In comparison to unlabelled cells in agarose medium, the labeled cells showed better visualization in the form of tiny distinct punctuate signal [151]. Chitosan composite can also be used in targeted drug delivery in combination with bioimaging. Fluorescent chitosan quantum dot composites were developed for targeted drug and gene delivery with simultaneous application of optical imaging [152,153].

2.8   Chitosan composite in cancer chemotherapy

Hydrophobically modified glycol chitosan (HGC) was prepared by conjugating it with 5β-cholanic acid employing 1-ethyl-3-(3-dimethylaminopropyl)-carbodiimide and then aniticancer drug paclitaxel was loaded into it by dialysis method. The resulted self-assembled NPs have the diameter of 400 nm and were found to be stable in phosphate buffer saline for 10 days. These NPs exhibited sustained drug release where 80% of loaded paclitaxel was released in 8 days at body temperature. Furthermore, the NPs were found to be less toxic to B16F10 melanoma cells compared to free paclitaxel prepared in cremophor EL injection [154,155]. In another study, glycol chitosan was conjugated with hydrophobic cholanic acid to form HGC. The resulted HGC converted into self-aggregates of nano-size in aqueous medium and then employed as carrier for cisplatin to treat cancers. The cisplatin loaded NPs of mean diameter about 300-500 nm and demonstrated sustained release of loaded cisplatin for a week. In addition, these NPs were less toxic than was free cisplatin, which may be due to sustained release of cisplatin from the NPs. Furthermore, the resulted NPs exhibited higher anti-tumour efficacy due to their higher accumulation in the tumour tissues, when tested in tumour bearing mice. Therefore, it was concluded that the developed NPs have promising ability to carry and release the loaded drug in target tissues [156].

Doxurubicin (DOX) is one of the widely used anticancer drugs. To counter its undesirable cardiotoxic effects, DOX was conjugated with dextran and then encapsulated in chitosan based hydrogel NPs. The resulted NPs of size 100 nm favours enhanced permeability and retention effect (EPR) as shown by majority of the solid tumours. These NPs demonstrated tumour regression and enhanced survival time compared to free drug as well as drug conjugate in antitumour study on mice. In addition, encapsulation of drug conjugates in chitosan NPs not only was able to reduce side effects, but also enhanced its therapeutic efficacy in the treatment of solid tumours [157]. In another study, DOX conjugated glycol-chitosan (GC-DOX) and fluorescein isothiocyanate (FITC) conjugated chitosan were developed to target tumor *in vivo*. Both the NPs were prepared by self-aggregation in aqueous media and the observed size of the resulted GC-DOX was 250-300 nm with loading content and loading efficiency were 38% and 97%, respectively. In the first step of *in vivo* study, FITC-GC NPs were injected into the tail vein of the tumour-induced rats and studied its distribution for 8 days. Thereafter, GC-DOX NPs were injected in to the tail vein of rats and their antitumour effect was investigated. It was observed that FITC-GC NPs were distributed in organs such as kidney, tumour tissues and liver and the level of distribution was maintained at a high level for 8 days with the increase in distribution in tumour tissues. This was attributed to EPR effect. The GC-DOX was able to suppress the tumour over 10 days [158].

Folate receptors are found in large numbers in retinoblastoma cells. Therefore, to reduce side effects and increase site specific drug delivery, DOX loaded chitosan NPs were conjugated with folic acid. The efficacy of the resulted conjugated DOX NPs on Y-79 retinoblast cells was investigated by MTT assay. The result showed that the conjugated NPs exhibited superior toxicity than the unconjugated NPs and native DOX. This may be due to higher (30%) intracellular uptake of conjugated NPs compared to 13.24% and 5.01% for unconjugated NPs and native DOX, respectively. The mechanism indicated that mitochondrial pathway was activated and the conjugated NPs were most effective in releasing cytochrome-C along with the activation of downstream caspases to help in the process of apoptosis. This resulted in a sustained and targeted delivery of DOX in tumour tissues [159]. Similarly, another conjugate composed of chitosan oligosaccharide and arachidic acid was successfully developed and used for the preparation of self-assembled NPs of DOX. The prepared NPs showed spherical shape with average diameter of 130 nm and positive surface charge. These NPs demonstrated sustained and pH dependent drug release profiles. In addition, the cellular uptake of DOX in FaDu cells (human head and neck cancerous cells) and anti-tumor efficacy in FaDu tumour xenografted mouse model of conjugated NPs-treated group were found to be higher compared to DOX treated group [160]. Table 1 presented different applications of chitosan composites with formulation types, composite materials with incorporated drugs.

Table 1. Various applications of chitosan composites with types of formulations, composite materials with incorporated drugs.

Types of formulation	Composite material (Excluding chitosan)	Drug	References
Chitosan composite in pharmaceutics/drug delivery			
Hydrogel microspheres	Polyacrylamide	Indomethacin	[34]
Microspheres	Poly(d,l-lactide-co-glycolide) (PLGA)	Deferoxamine	[35]
Composite nanoparticles	Poly (vinyl alcohol) (PVA) and Magnetic particles (Fe_3O_4)	Bovine serum albumin (BSA)	[36]
Composite microspheres	Ethyl cellulose	Rifabutin	[39]
Chitosan film incorporated with Microspheres	Ethyl cellulose	Ciprofloxacin HCl release	[42]
Composite sponges	HPMC	Tripterine	[43]
Composite microspheres	Poly(N-isopropylacrylamide) (PNIPAM)) and Magnetic particles	Indomethacin	[44]
Composite microspheres	PNIPAM	Berberine HCl	[45]
Composite nanofibers	PLGA and polyuratheane (PU)	Danshensu	[46]
Composite microparticles	Silicon particles	BSA & Insulin	[47]

Chitosan composite in gene delivery			
composite hydrogel	Pluronic	Gene therapy	[53]
Chitosan composite in tissue enginerring			
Sponges	Beta-tricalcium phosphate and CP	Gentamycin	[68]
Scaffold	Hydroxyapatite or CP glass	----	[69]
Scaffold	Hydroxyapatite and gelatin	----	[71]
Composite bone cement	Hydroxyapatite and polymethylmethacrylate (PMMA)	----	[72]
Scaffold	Polyglactin fibers and CPC	Mesenchymal stem cells obtained from human umbilical cord	[73]
Biocomposite	Hydroxyapatite, titania powder and gelatin	--------	[77]
3D hybrid scaffolds with micro-porous structure	N-(2-carboxybenzyl)chitosan and Titania nanoparticles and bioglass 45S5	-------	[78]
Nano-composite scaffold	Gelatin and bioactive glass ceramic nanoparticles	--------	[79]
Nanocomposite	Carbon nanotubes (f-MWCNT)	-------	[80]
Scaffold	Hydroxyethyl chitosan and cellulose	-------	[82]
Injectable composite hydrogels	Methacrylated glycol chitosan and hyaluronic acid	Riboflavin	[84]
Injectable composite	carboxymethyl chitosan (CMC) and collagen nanofibers	-----	[88]
Composite hydrogel	Carrageenan	------	[94]
Composite matrix	Alginate and hyaluronan	-------	[95]
Composite microparticles	Silk matrix		[96]
Semi-interpenetrating polymer network (IPN) scaffold	Alginate	--------	[97]
Composite membrane	Poly(l-lysine)	----	[102]
Composite membrane	poly(L-lysine), polyethyleneimine and poly-L-ornithine	----	
Composite films	Gelatin	------	[104]
Nano fibrous scaffold	Poly(3-hydroxybutyrate)	-----	[105]
Nerve conduits	PDLLA and chondroitin sulfate	-----	[106]
Thermosensitive hydrogels	Gelatin	Adipose-derived stem cells	[109]
Scaffold	Carbon nanofibers	-------	[112]

Scaffold	Graphene oxide gold nanosheets (GO-Au)	-----	[113]
Polymeric composites	Collagen	-----	[115]
Scaffold	Galatosylated chitosan and oxidized alginate	------	[116]
Sponge	Galactosylated chitosan and alginate	------	[117]
Scaffold	Silk fibroin	-------	[118]
Scaffold	Silk fibroin	-------	[119]
Chitosan composite in preparation of implants			
Implant	Ag-NPs	-----	[122]
Implatable tablets	Polyvinyl chloride (PVC)	Risedronate sodium	[123]
Nanocomposite implants	Carbon nanotubes (CNT) and hydroxyapatite	-------	[124]
Implatable pellets	Bovine hydroxyapatite	Ciprofloxacin	[125]
Chitosan composite in the preparation of contact lenses			
Composite films	Gelatin	-----	[127]
Chitosan composite in wound healing			
Hydrogel scaffold	Ag nanoparticles	----	[134]
Scaffold	Ag nanoparticles	----	[135]
Composite sponges	Succinyl pullulan-carboxymethyl chitosan	-----	[138]
Composite hydrogel	Zinc oxide NPs	----	[139]
Composite hydrogel	CMC and gelatin	----	[140]
Composite hydrogel	N, O-CMC and oxidized alginate	Curcumin	[141]
Injectable hydrogel	Alginate	Nano-curcumin	[142]
Hydrogel	PVA	Minocycline	[143]
Nano-hydrogel	Platelet rich plasma	Tigecycline	[144]
Composite film	Thiolated chitosan and poly (N-isopropyl acrylamide)	Ciprofloxacin	[147]
Composite film	PVP and TiO_2	-----	[148]
Chitosan composite in bioimaging			
Self-assembled nanoparticles	Fe_3O_4, Linoleic acid	----	[150]
Coated Particles	CMC and superparamagnetic iron oxide	----	[151]
Chitosan composite in cancer chemotherapy			
Self-assembled nanoparticles	Glycol chitosan (HGC) and 5β-cholanic acid employing 1-ethyl-3-(3-dimethylaminopropyl)-carbodiimide	Paclitaxel	[154,155]
Nano-size self-aggregates	Glycol chitosan and cholanic acid	Cisplatin	[156]

Hydrogel nanoparticles	Dextran	Doxurubicin	[157]
Conjugated nanoparticles	Glycol-chitosan (GC-DOX) and fluorescein isothiocyanate (FITC)	Doxurubicin	[158]
Self-assembled nanoparticles	Chitosan oligosaccharide and arachidic acid	Doxurubicin	[160]

3. Applications of chitosan composites in food sectors

Biodegradable materials based active food packaging is the need of the hour in order to fulfil the demand of consumers towards food safety and environment related issues. In this context, some of the carbohydrate polymers including chitosan, starch, cellulose derivatives and pectins are widely used as packaging material not only to provide biodegradable but also to inculcate edible property [161]. Among the above biopolymers, chitosan is an exciting and promising one due to its additional property such as antimicrobial and antioxidant properties which increase the shelf-life of packaged food and its film forming ability [162]. However, application of pure chitosan film suffers due to brittleness and limited antioxidant and antimicrobial properties. Therefore, to improve mechanical and biological properties of chitosan film it is essential to incorporate another antioxidant and/or antimicrobial ingredient into it [163]. Among different kinds of natural products, phenolic compounds are widely available in nature because of their biological activities such as antioxidant, antimicrobial anti-inflammatory, anti-diabetic, and anticancer [164]. The antioxidant properties of the polyphenols are due to their metal chelating property, inhibition of lipooxygenase activity and free radical scavenging properties [165].

The influence of phenolic extract obtained from seedcake of *Lepidium sativum* (LSE) on the structural, mechanical and functional attributes of chitosan film was investigated. The resulted composite films demonstrated structural modification, which was confirmed by Fourier-transform infrared spectroscopy (FTIR) study. However, the film surface was observed to be smooth and homogenous. Incorporation of 5% v/v of LSE resulted in the improvement of tensile strength of 32.2% and elongation of 109% of the chitosan film in contrast to tannic acid as reference. This is attributed to the decrease in free space which hinders the molecular movement within the developed film. The resulted film demonstrated adequate release of phenolic compound in dose and time dependent way. The antioxidant activity of the resulted film in terms of free radical scavenging activity (2,2-diphenyl-1-picrylhydrazyl, DPPH) and frerric reducing ability power (FRAP) activity followed the same pattern in both aqueous and fatty food stimulants. It was concluded that LSE could be used in the development of novel bioactive and biodegradable composite food packaging film for food preservation with chitosan [166].

By incorporating extract of sweat potato (*Ipomoea batatas L.*) into the chitosan matrix an intelligent pH-sensing films with antioxidant property were developed. With the increase in extract amount, properties of films such as antioxidant, thermal stability, light blocking attributes and pH sensing ability were increased. However, the films exhibited a decrease in water vapour barrier property and tensile strength with increase in extract content in the films. Microscopic observation showed prepared films are free from non-homogeneity and FTIR study confirmed that the presence of intermolecular intractions between constituents of extract and chitosan matrix. The resulted film with 10 wt% of extract exhibited highest free radical scavenging properties in time dependent manner. Additionally, the films based on chitosan and extract showed distinct colour changes along with the pH change. Consequently, the resulted film could be an excellent packaging material to prolong the self-life and to observe the quality of food products [167].

Novel packaging films were developed with the inclusion of protocatechuic acid (PA) in chitosan matrix and the resulted films were studied for various properties such as physical, mechanical, structural and antioxidant. The above attributes of the chitosan-PA based films intimately related with the content of PA in the film. The lower amount (<1%) of PA had increased the tensile strength and reduces water vapour permeability due to reduction in free space and molecular movement within film. However, higher content of PA (>1.5%) lessen the homogeneity and raising the crystallinity potential of film resulting in higher free volume and molecular movement within the film. Therefore, above composite films were capable of releasing phenolic compounds leading to antioxidant potential for aqueous and fatty food materials. Furthermore, films having either 0.5% or 1% of PA are considered for the preparation of final active packaging material for food preservation [168]. In another study, PA was grafted into chitosan matrix film at different ratios and then characterized for physiochemical, mechanical and antioxidant properties. The resulted films were found to be transparent and the thickness was ranging from 44.1 to 48.6 mm. The film showed increased water solubility and tensile strength. However, moisture content and water vapour permeability were found to be reduced. Antioxidant activity study using DPPH radical method demonstrated that the resulted composite film had both time-dependent and dose-dependent radical scavenging activity and the composite films could be used as novel antioxidant film for food packing material [169].

Mango leaf extract (1-5%) was incorporated into chitosan films and studied for its effect on morphology, mechanical property, moisture and oxygen permeability, and antioxidant property. The composite films were found to be compact, smooth and dense nature. With the inclusion of extract resulted in lower water vapour and oxygen permeability. The tensile strength was found to be higher whereas elongation at break showed a reduction compared to chitosan film alone. Antioxidant studies in terms of DPPH radical scavenging, total phenolic content and ferric reducing power demonstrated increased antioxidant activity with the incremental amounts of extract inclusion in the chitosan matrix. In addition, mango leaf extract inclusion in the chitosan matrix preserved the cashew nuts with 56% higher oxidation resistance for the 5% extract containing film compared to commercial polyamide and polyethylene film [170]. Chitosan film incorporated with apple peel polyphenol demonstrated improved various physical attributes of the film. However, water vapour permeability and moisture content were observed to be decreased. In addition, tensile strength and elongation at break of the composite film were significantly lower than those of the film containing chitosan alone. Moreover, the antioxidant and antimicrobial ability of the composite film were remarkably enhanced and composite film containing 0.5% of apple peel polyphenol showed better properties in terms of above parameters [171].

Composite films of chitosan and starch blended with polyphenol obtained from thyme extract were developed and the release of polyphenol from the composite films was studied in different media. The release study in aqueous medium showed that the release rate of polyphenol was less compared to starch film. The above result was due to strong interaction between chitosan and polyphenol. However, higher polyphenol release was observed in acidic medium which is due to higher solubility of chitosan in the medium. The incorporation of tannic acid into the composite films enhanced the cross-linking process leading to delay in polyphenol release. Finally, it was concluded that the polyphenol-chitosan matrix interactions significantly influenced the polyphenol release and subsequently the antioxidant activity of the composite films [172].

Conclusions

This chapter discusses various applications of chitosan composites in pharmaceutical and food packaging sectors. Chitosan is an excellent biopolymer having desired pharmaceutical attributes such as mucoadhesive, biodegradability, biocompatibility, and swelling. Therefore, it is being widely used in delivering the drugs of diverse classes. In spite of all, chitosan has limitations including low aqueous solubility, poor mechanical strength and thermal instability. The most effective way to dealt with this is to transform chitosan into composites. This conversation is viable because of the existence of ammonia and hydroxyl groups on chitosan structure. This led to the use of chitosan composites not only in pharmaceutical but also in biomedical and food sectors. So far, no data is available on toxicological aspects of chitosan composites. Therefore, pre-clinical investigation on chitosan as well as on its derivatives would establish its safety profile.

References

[1] M. Fazeli, J. Florez, R. Simão, Improvement in adhesion of cellulose fibers to the thermoplastic starch matrix by plasma treatment modification, Composites Part B: Eng. 163 (2018) 207-216. https://doi.org/10.1016/j.compositesb.2018.11.048

[2] Y. Shi, S. Jiang, K. Zhou, C. Bao, B. Yu, X. Qian, B. Wang, N. Hong, P. Wen, Z. Gui, Y. Hu, R.K. Yuen, Influence of g-C3N4 nanosheets on thermal stability and mechanical properties of biopolymer electrolyte nanocomposite films: A novel investigation, ACS Appl. Mater. Interfac. 6 (2014) 429-437. https://doi.org/10.1021/am4044932

[3] Y.X. Xu, K.M. Kim, M.A. Hanna, D. Nag, Chitosan-starch composite film: Preparation and characterization, Ind. Crops Prod. 21 (2005) 185-192. https://doi.org/10.1016/j.indcrop.2004.03.002

[4] X. Wang, B. Xing, Importance of structural makeup of biopolymers for organic contaminant sorption, Environ. Sci. Technol. 41 (2007) 3559–3565. https://doi.org/10.1021/es062589t

[5] J.K. Park, M.J. Chung, H.N. Choi, Y.I. Park, Effects of the molecular weight and the degree of deacetylation of chitosan oligosaccharides on antitumor activity, Int. J. Mol. Sci. 12 (2011) 266–277. https://doi.org/10.3390/ijms12010266

[6] Y.S. Puvvada, S. Vankayalapati, S. Sukhavasi, Extraction of chitin from chitosan from exoskeleton of shrimp for application in the pharmaceutical industry, Int. Curr. Pharm. J. 1 (2012) 258–263. https://doi.org/10.3329/icpj.v1i9.11616

[7] E. Khor, L.Y. Lim, Implantable applications of chitin and chitosan, Biomaterials. 24 (2003) 2339–2349. https://doi.org/10.1016/S0142-9612(03)00026-7

[8] R. Jayakumar, K.P. Chennazhi, S. Srinivasan, S.V. Nair, T. Furuike, H. Tamura, Chitin scaffolds in tissue engineering, Int. J. Mol. Sci. 12 (2011) 1876–1887. https://doi.org/10.3390/ijms12031876

[9] I.Y. Kim, S.J. Seo, H.S. Moon, M.K. Yoo, I.Y. Park, B.C. Kim, C.S. Cho, Chitosan and its derivatives for tissue engineering applications, Biotechnol Adv. 26 (2008) 1-21. https://doi.org/10.1016/j.biotechadv.2007.07.009

[10] M. Rinaudo, Chitin and chitosan: properties and applications, Prog. Polym. Sci. 31 (2006) 603–632. https://doi.org/10.1016/j.progpolymsci.2006.06.001

[11] K. Kiene, F. Porta, B. Topacogullari, P. Detampel, J. Huwyler, Self-assembling chitosan hydrogel: A drug-delivery device enabling the sustained release of proteins, J. Appl. Polym. Sci. 135(1) (2018) 45638. https://doi.org/10.1002/app.45638

[12]   M.C.G. Pellá, M.K. Lima-Tenório, E.T. Tenório-Neto, M.R. Guilherme, E.C. Muniz, A.F. Rubira, Chitosan-based hydrogels: From preparation to biomedical applications, Carbohydr. Polym. 196 (2018) 233–245. https://doi.org/10.1016/j.carbpol.2018.05.033

[13]   P. Sacco, M. Borgogna, A. Travan, E. Marsich, S. Paoletti, F. Asaro, I. Donati, Polysaccharide-based networks from homogeneous chitosan-tripolyphosphate hydrogels: Synthesis and characterization, Biomacromol. 15 (2014) 3396–3405. https://doi.org/10.1021/bm500909n

[14]   P. Sacco, S. Paoletti, M. Cok, F. Asaro, M. Abrami, M. Grassi, I. Donati, Insight into the ionotropic gelation of chitosan using tripolyphosphate and pyrophosphate as cross-linkers, Int. J. Biol. Macromol. 92 (2016) 476–483. https://doi.org/10.1016/j.ijbiomac.2016.07.056

[15]   J. Venkatesan, P.A. Vinodhini, P.N. Sudha, S.K. Kim, Chitin and chitosan composites for bone tissue regeneration, Adv. Food Nutr. Res. 73 (2014) 59–81. https://doi.org/10.1016/B978-0-12-800268-1.00005-6

[16]   L. Tan, J. Hu, H. Huang, J. Han, H. Hu, Study of multi-functional electrospun composite nanofibrous mats for smart wound healing. Int. J. Biol. Macromol. 79 (2015) 469–476. https://doi.org/10.1016/j.ijbiomac.2015.05.014

[17]   J.K. Vasir, K. Tambwekar, S. Garg, Bioadhesive microspheres as a controlled drug delivery system, Int. J. Pharm. 255 (2003) 13–32. https://doi.org/10.1016/S0378-5173(03)00087-5

[18]   K. Vimala, Y.M. Mohan, K. Varaprasad, N.N. Redd, S. Ravindra, N.S. Naidu, K.M. Raju, Fabrication of curcumin encapsulated chitosan-PVA silver nanocomposite films for improved antimicrobial activity, J. Biomater. Nanobiotechnol. 2 (2011) 55–64. https://doi.org/10.4236/jbnb.2011.21008

[19]   K. Kofuji, C.J. Qian, M. Nishimura, I. Sugiyama, Y. Murata, S. Kawashima, Relationship between physicochemical characteristics and functional properties of chitosan, Eur. Polym. J. 41 (2005) 2784–2791. https://doi.org/10.1016/j.eurpolymj.2005.04.041

[20]   W. Sun, S. Mao, Y. Wang, V.B. Junyaprasert, T. Zhang, L. Na, J. Wang, Bioadhesion and oral absorption of enoxaparin nanocomplexes, Int. J. Pharm. 386 (2010) 275-281. https://doi.org/10.1016/j.ijpharm.2009.11.025

[21]   G. Shavi, U. Nayak, M. Reddy, A. Karthik, P.B. Deshpande, A.R. Kumar, N. Udupa, Sustained release optimized formulation of anastrozole-loaded chitosan

microspheres: In vitro and in vivo evaluation, Mater. Sci. Mater. Med. 22 (2011) 865-878. https://doi.org/10.1007/s10856-011-4274-y

[22]   E. Meng-Lund, C. Muff-Westergaard, C. Sander, P. Madelung, J. Jacobsen, A mechanistic based approach for enhancing buccal mucoadhesion of chitosan. Int J Pharm. 461 (2014) 280–285. https://doi.org/10.1016/j.ijpharm.2013.10.047

[23]   A. Jintapattanakit, V.B. Junyaprasert, T.J. Kissel, The role of mucoadhesion of trimethyl chitosan and PEGylated trimethyl chitosan nanocomplexes in insulin uptake, Pharm. Sci. 98 (2009) 4818-30. https://doi.org/10.1002/jps.21783

[24]   Saikia C, Gogoi P, T.K. Maji, Chitosan: a promising biopolymer in drug delivery applications, J. Mol. Genet. Med. 6 (2015) 1-10. https://doi.org/10.4172/1747-0862.S4-006

[25]   R. Parhi, Chitin and Chitosan in Drug Delivery, in: G. Crini, E. Lichtfouse, (Eds.), Sustainable Agriculture Reviews, Springer, Cham, 2019, pp. 175-239. https://doi.org/10.1007/978-3-030-16581-9_6

[26]   S. Gupta, S.P. Vyas, Carbopol/chitosan based pH triggered in situ gelling system for ocular delivery of timolol maleate, Sci. Pharm. 78 (2010) 959-76. https://doi.org/10.3797/scipharm.1001-06

[27]   D. Sakloetsakun, J. Hombach, A. Bernkop-Schnurch, In situ gelling properties of chitosan ethioglycolic acid conjugate in the presence of oxidizing agents, Biomater. 30 (2009) 6151-7. https://doi.org/10.1016/j.biomaterials.2009.07.060

[28]   J. Malmo, K.M. Vrum, S.P. Strand, Effect of chitosan chain architecture on gene delivery: Comparison of self-branched and linear chitosans, Biomacromol. 12 (2011) 721-9. https://doi.org/10.1021/bm1013525

[29]   R. Martien, B. Loretz, M. Thaler, S. Majzoob, A. Bernkop-Schnurch, Chitosanethioglycolic acid conjugate: An alternative carrier for oral nonviral gene delivery, J. Biomed. Mater. Res. A. 82 (2007) 1-9. https://doi.org/10.1002/jbm.a.31135

[30]   M. Malhotra, C. Lane, C. Tomaro-Duchesneau, S. Saha, S. Prakash, A novel scheme for synthesis of PEG-graftedchitosan polymer for preparation of nanoparticles and other applications, Int. J. Nanomed. 6 (2011) 485-94.

[31]   N.G.M. Schipper, S. Olsson, J.A. Hoogstraate, A.G. deBoer, K.M. Va°rum, P. Artursson, Chitosans as absorption enhancer for poorly absorbable drugs: 2. Mechanism of absorption enhancement, Pharm. Res. 14 (1997) 923-9. https://doi.org/10.1023/A:1012160102740

[32]   P. Shah, V. Jogani, P. Mishra, A.K. Mishra, T. Bagchi, A. Misra, Modulation of ganciclovir intestinal absorption in presence of absorption enhancers, J. Pharm. Sci. 96 (2007) 2710-22. https://doi.org/10.1002/jps.20888

[33]   C.E. Kast, A. Bernkop-Schnurch, Influence of the molecular mass on the permeation enhancing effect of different poly(acrylates), STP Pharm. Sci. 6 (2002) 351-6.

[34]   L. Zhang, Y. Ma, X. Pan, S. Chen, H. Zhuang, S. Wang, A composite hydrogel of chitosan/heparin/poly (γ-glutamic acid) loaded with superoxide dismutase for wound healing, Carbohydrate Polym. 180 (2018) 168–174. https://doi.org/10.1016/j.carbpol.2017.10.036

[35]   G. Rassu, A. Salis, E.P. Porcu, P. Giunchedi, M. Roldo, E. Gavini, Composite chitosan/alginate hydrogel for controlled release of deferoxamine: A system to potentially treat iron dysregulation diseases, Carbohydr. Polym. 136 (2016) 1338–1347. https://doi.org/10.1016/j.carbpol.2015.10.048

[36]   E.T. Nicknejad, S.M. Ghoreishi, N. Habibi, Electrospinning of cross-linked magnetic chitosan nanofibers for protein release, AAPS Pharm Sci Tech. 16 (2015) 1480-86. https://doi.org/10.1208/s12249-015-0336-7

[37]   R. Sharma, D. Saxena, A.K. Dwivedi, A. Misra, Inhalable microparticles containing drug combinations to target alveolar macrophages for treatment of pulmonary tuberculosis, Pharm. Res. 18 (2001) 1405–1410. https://doi.org/10.1023/A:1012296604685

[38]   K. Hirota, T. Hasegawa, T. Nakajima, H. Inagawa, C. Kohchi, G. Soma, K. Makino, H. Terada, Delivery of rifampicin-PLGA microspheres into alveolar macrophages is promising for treatment of tuberculosis, J. Control. Release. 142 (2010) 339–346. https://doi.org/10.1016/j.jconrel.2009.11.020

[39]   H. Feng, L. Zhang, C. Zhu, Genipin crosslinked ethyl cellulose–chitosan complex microspheres for anti-tuberculosis delivery, Colloid Surf. B. Biointerf. 103 (2013) 530–537. https://doi.org/10.1016/j.colsurfb.2012.11.007

[40]   H. Nazar, D.G. Fatouros, S.M. van der Merwe, N. Bouropoulos, G. Avgouropoulos, J. Tsibouklis, M. Roldo, Thermosensitive hydrogels for nasal drug delivery: the formulation and characterisation of systems based on N-trimethyl chitosan chloride, Eur. J. Pharm. Biopharm. 77 (2011) 225-32. https://doi.org/10.1016/j.ejpb.2010.11.022

[41]   Z. Liang, T. Gong, X. Sun, J.Z. Tang, Z. Zhang, Chitosan oligomers as drug carriers for renal delivery of zidovudine, Carbohydr. Polym. 87 (2012) 2284-90. https://doi.org/10.1016/j.carbpol.2011.10.060

[42]   P. Shi, Y. Zuo, Q. Zou, J. Shen, L. Zhang, Y. Li, Y.S. Morsi, Improved properties of incorporated chitosan film with ethyl cellulose microspheres for controlled release, Int. J. Pharm. 375 (2009) 67–74. https://doi.org/10.1016/j.ijpharm.2009.04.016

[43]   M.S. Freag, W.M. Saleh, O.Y. Abdallah, Laminated chitosan-based composite sponges for transmucosal delivery of novel protamine-decorated tripterine phytosomes: Ex-vivo mucopenetration and in-vivo pharmacokinetic assessments, Carbohydr, Polym. 188 (2018) 108–120. https://doi.org/10.1016/j.carbpol.2018.01.095

[44]   N. Rodkate, M. Rutnakornpituk, Multi-responsive magnetic microsphere of poly(N-isopropylacrylamide)/carboxymethylchitosan hydrogel for drug controlled release, Carbohydr. Polym. 151 (2016) 251–259. https://doi.org/10.1016/j.carbpol.2016.05.081

[45]   Y. Gong, Q.L. Liu, A.M. Zhu, Q.G. Zhang, One-pot synthesis of poly(N-isopropylacrylamide)/chitosan composite microspheres via microemulsion, Carbohydr. Polym. 90 (2012) 690–695. https://doi.org/10.1016/j.carbpol.2012.05.098

[46]   H. Yu, T. Zhu, J. Xie, J. Du, C. Sun, J. Wang, J. Wang, S. Chen, Preparation of inorganic-organic-framework nanoscale carries as a potential platform for drug delivery, Adv. Eng. Mater. 1800626 (2018) 1-9. https://doi.org/10.1002/adem.201800626

[47]   E. Pastor, E. Matveeva, A. Valle-Gallego, F.M. Goycoolea, M. Garcia-Fuentes, Protein delivery based on uncoated and chitosan-coated mesoporous silicon microparticles, Colloid Surf. B Biointerf. 88 (2011) 601–609. https://doi.org/10.1016/j.colsurfb.2011.07.049

[48]   M. Prabaharan, J.F. Mano, Chitosan-based particles as controlled drug delivery systems, Drug Deliv. 12 (2005) 41–57. https://doi.org/10.1080/10717540590889781

[49]   S. Mao, W. Sun, T. Kissel, Chitosan-based formulations for delivery of DNA and siRNA, Adv. Drug Deliv. Rev. 62 (2009) 12–27. https://doi.org/10.1016/j.addr.2009.08.004

[50]   T. Sato, T. Ishii, Y. Okahata, In vitro gene delivery mediated by chitosan. Effect of pH, serum, and molecular mass of chitosan on the transfection efficiency, Biomater. 22 (2001) 2075–80. https://doi.org/10.1016/S0142-9612(00)00385-9

[51] J.I. Lee, H.S. Kim, H.S. Yoo, DNA nanogels composed of chitosan and pluronic with thermo-sensitive and photo-crosslinking properties, Int. J. Pharm. 373 (2009) 93–9. https://doi.org/10.1016/j.ijpharm.2009.01.016

[52] O. Ortona, G. D'Errico, G. Mangiapia, D .Ciccarelli, The aggregative behavior of hydrophobically modified chitosans with high substitution degree in aqueous solution, Carbohydr. Polym. 74 (2008) 16–22. https://doi.org/10.1016/j.carbpol.2008.01.009

[53] Y.Z. Du, L. Wang, H. Yuan, F.Q. Hu, Linoleic acid-grafted chitosan oligosaccharide micelles for intracellular drug delivery and reverse drug resistance of tumor cells, Int. J. Biol. Macromol. 48 (2011) 215–222. https://doi.org/10.1016/j.ijbiomac.2010.11.005

[54] S. Ahmed, A.A. Ali, J. Sheikh, A review on chitosan centred scaffolds and their applications in tissue engineering, Int. J. Biol. Macromol. 116 (2018) 849–862. https://doi.org/10.1016/j.ijbiomac.2018.04.176

[55] T.H. Qazi, R. Rai, A.R. Boccaccini, Tissue engineering of electrically responsive tissues using polyaniline based polymers: A review, Biomater. 35 (33) (2014) 9068–9086. https://doi.org/10.1016/j.biomaterials.2014.07.020

[56] H. Mittal, S.S. Raya, B.S. Kaithd, J.K. Bhatiad, Sukritid, J. Sharmad, S.M. Alhassan, Recent progress in the structural modification of chitosan for applications in diversified biomedical fields, Eur. Polym. J. 109 (2018) 402–434. https://doi.org/10.1016/j.eurpolymj.2018.10.013

[57] R.C.F. Cheung, T.B. Ng, J.H. Wong, W.Y. Chan, Chitosan: An Update on Potential Biomedical and Pharmaceutical Applications, Mar. Drugs. 13 (2015) 5156-5186. https://doi.org/10.3390/md13085156

[58] S.D. Baljinder, B.A. Adetola, Current clinical therapies for cartilage repair, their limitation and the role of stem cells, Curr. Stem Cell Res. Ther. 7 (2012) 143–148. https://doi.org/10.2174/157488812799219009

[59] C. Chai, K.W. Leong, Biomaterials approach to expand and direct differentiation of stem cells, Mol. Ther. 15 (2007) 467–480. https://doi.org/10.1038/sj.mt.6300084

[60] Y.J. Seol, J.-Y. Lee, Y.J. Park, Y.-M. Lee, Y. Ku, I.C. Rhyu, S.J. Lee, S.B. Han, C.-P. Chung, Chitosan sponges as tissue engineering scaffolds for bone formation, Biotech. Lett. 26 (2004) 1037–1041. https://doi.org/10.1023/B:BILE.0000032962.79531.fd

[61]   J.H. Jang, O. Castano, H.-W. Kim, Electrospun materials as potential platforms for bone tissue engineering, Adv. Drug Deliv. Rev. 61 (2009) 1065–1083. https://doi.org/10.1016/j.addr.2009.07.008

[62]   I. Cacciotti, Cationic and anionic substitutions in hydroxyapatite, in: I.V. Antoniac (Ed.), Handbook of Bioceramics and Biocomposites, Springer International Publishing, Cham, 2016, pp. 145–211. https://doi.org/10.1007/978-3-319-12460-5_7

[63]   S.J. Hollister, Porous scaffold design for tissue engineering, Nat. Mater. 4 (2005) 518–524. https://doi.org/10.1038/nmat1421

[64]   H. Seeherman, R. Li, J. Wozney, A review of preclinical program development for evaluating injectable carriers for osteogenic factors, J. Bone Joint Surg. Am. 85A(Suppl 3) (2003) 96–108. https://doi.org/10.2106/00004623-200300003-00016

[65]   Y. Zhang, M. Zhang, Synthesis and characterization of macroporous chitosan/ calcium phosphate composite scaffolds for tissue engineering, J. Biomed. Mater. Res. 55 (2001) 304–312. https://doi.org/10.1002/1097-4636(20010605)55:3<304::AID-JBM1018>3.0.CO;2-J

[66]   S. Deepthi, J. Venkatesan, Se-Kwon Kim, Joel D. Bumgardner, R. Jayakumar, An overview of chitin or chitosan/nano ceramic composite scaffolds for bone tissue engineering, Int. J. Biol. Macromol. 93 (2016) 1338–1353. https://doi.org/10.1016/j.ijbiomac.2016.03.041

[67]   Y. Zhang, M. Zhang, Three-dimensional macroporous calcium phosphate bioceramics with nested chitosan sponges for loadbearing bone implants, J. Biomed. Mater. Res. 61 (2002) 1–8. https://doi.org/10.1002/jbm.10176

[68]   Y. Zhang, M. Zhang, Calcium phosphate/chitosan composite scaffolds for controlled in vitro antibiotic drug release, J. Biomed. Mater. Res. 62 (2002) 378–86. https://doi.org/10.1002/jbm.10312

[69]   Y. Zhang, M. Ni, M. Zhang, B. Ratner, Calcium phosphatechitosan composite scaffolds for bone tissue engineering, Tissue Eng. 9 (2003) 337–45. https://doi.org/10.1089/107632703764664800

[70]   H.H. Xu, J.B. Quinn, S. Takagi, L.C. Chow, Synergistic reinforcement of in situ hardening calcium phosphate composite scaffold for bone tissue engineering, Biomater. 25 (2004) 1029–37. https://doi.org/10.1016/S0142-9612(03)00608-2

[71]   F. Zhao, Y. Yin, W.W. Lu, J.C. Leong, W. Zhang, J. Zhang, M. Zhang, K. Yao, Preparation and histological evaluation of biomimetic threedimensional

hydroxyapatite/chitosan-gelatin network composite scaffolds, Biomater. 23 (2002) 3227–34. https://doi.org/10.1016/S0142-9612(02)00077-7

[72]   S.B. Kim, Y.J. Kim, T.L. Yoon, S.A. Park, I.H. Cho, E.J. Kim, I.A. Kim, J.W. Shin, The characteristics of a hydroxyapatite-chitosan-PMMA bone cement, Biomater. 25 (2004) 5715–23 https://doi.org/10.1016/j.biomaterials.2004.01.022

[73]   L. Zhao, E.F. Burguera, H.H. Xu, N. Amin, H. Ryou, D.D. Arola, Fatigue and human umbilical cord stem cell seeding characteristics of calcium phosphate-chitosan-biodegradable fiber scaffolds, Biomater. 31 (2010) 840–7. https://doi.org/10.1016/j.biomaterials.2009.09.106

[74]   C. Chang, N. Peng, M. He, Y. Teramoto, Y. Nishio, L. Zhang, Fabrication and properties of chitin/hydroxyapatite hybrid hydrogels as scaffold nano-materials, Carbohydr. Polym. 91 (2013) 7–13 https://doi.org/10.1016/j.carbpol.2012.07.070

[75]   S.B. Qasim, S. Husain, Y. Huang, M. Pogorielov, V. Deineka, M. Lyndin, A. Rawlinson, I.U. Rehman, In-vitro and in-vivo degradation studies of freeze gelated porous chitosan composite scaffolds for tissue engineering applications, Polym. Degrad. Stab. 136 (2017) 31–38. https://doi.org/10.1016/j.polymdegradstab.2016.11.018

[76]   A. Olad, F. Farshi Azhar, The synergetic effect of bioactive ceramic and nanoclay on the properties of chitosan–gelatin/nanohydroxyapatite–montmorillonite scaffold for bone tissue engineering, Ceram. Int. 40 (7 Part A) (2014) 10061–10072. https://doi.org/10.1016/j.ceramint.2014.04.010

[77]   K.R. Mohamed, A.A. Mostafa, Preparation and bioactivity evaluation of hydroxyapatite-titania/chitosan-gelatin polymeric biocomposites, Mater. Sci. Eng. C. 28(7) (2008) 1087–1099. https://doi.org/10.1016/j.msec.2007.04.040

[78]   M. Nerantzaki, M. Nerantzaki, I.G. Koliakou, M. Kaloyianni, M. Kaloyianni, D.N. Bikiari, D.N. Bikiaris, New N-(2-carboxybenzyl)chitosan composite scaffolds containing nanoTiO$_2$ or bioglass with enhanced cell proliferation for bone-tissue engineering applications, Int. J. Polym. Mater. 66(2) (2016) 71-81. https://doi.org/10.1080/00914037.2016.1182913

[79]   M. Peter, N.S. Binulal, S.V. Nair, N. Selvamurugan, H. Tamura, R. Jayakumar, Novel biodegradable chitosan–gelatin/nano-bioactive glass ceramic composite scaffolds for alveolar bone tissue engineering, Chem. Eng. J. 158 (2010) 353–361. https://doi.org/10.1016/j.cej.2010.02.003

[80]   J. Venkatesan, B. Ryu, P.N. Sudha, S.K. Kim, Preparation and characterization of chitosancarbon nanotube scaffolds for bone tissue engineering, Int. J. Biol. Macromol. 50 (2012) 393–402. https://doi.org/10.1016/j.ijbiomac.2011.12.032

[81]   L. Yang, Q. Wang, L. Peng, H. Yue, Z. Zhang, Vascularization of repaired limb bone defects using chitosan-β-tricalcium phosphate composite as a tissue engineering bone scaffold, Mol. Med. Rep. 12 (2015) 2343–2347. https://doi.org/10.1016/j.ijbiomac.2011.12.032

[82]   Y. Wang, J. Qian, N. Zhao, T. Liu, W. Xu, A. Suo, Novel hydroxyethyl chitosan/cellulose scaffolds with bubble-like porous structure for bone tissue engineering, Carbohydr. Polym. 167 (2017) 44–51. https://doi.org/10.1016/j.carbpol.2017.03.030

[83]   M.L. Alves da Silva, A. Crawford, J.M. Mundy, V.M. Correlo, P. Sol, M. Bhattacharya, P.V. Hatton, R.L. Reis, N.M. Neves, Chitosan/polyester-based scaffolds for cartilage tissue engineering: Assessment of extracellular matrix formation, Acta Biomater. 6 (2010) 1149–1157. https://doi.org/10.1016/j.actbio.2009.09.006

[84]   H. Park, B. Choi, J. Hu, M. Lee, Injectable chitosan hyaluronic acid hydrogels for cartilage tissue engineering, Acta Biomater. 9 (2013) 4779–4786. https://doi.org/10.1016/j.actbio.2012.08.033

[85]   W.A. Li, B.Y. Lu, L. Gu, Y. Choi, J. Kim, D.J. Mooney, The effect of surface modification of mesoporous silica micro-rod scaffold on immune cell activation and infiltration, Biomater. 83 (2016) 249–256. https://doi.org/10.1016/j.biomaterials.2016.01.026

[86]   C.Y. Kuo, C.H. Chen, C.Y. Hsiao, J.P. Chen, Incorporation of chitosan in biomimetic gelatin/chondroitin-6-sulfate/hyaluronan cryogel for cartilage tissue engineering, Carbohydr. Polym. 117 (2015) 722–730. https://doi.org/10.1016/j.carbpol.2014.10.056

[87]   N.S. Sambudi, M. Sathyamurthy, G.M. Lee, S.B. Park, Electrospun chitosan/poly (vinyl alcohol) reinforced with $CaCO_3$ nanoparticles with enhanced mechanical properties and biocompatibility for cartilage tissue engineering, Compos. Sci. Technol. 106 (2015) 76–84. https://doi.org/10.1016/j.compscitech.2014.11.003

[88]   G. Lu, B. Sheng, Y. Wei, G. Wang, L. Zhang, Q. Ao, Y. Gong, X. Zhang, Collagen nanofiber-covered porous biodegradable carboxymethyl chitosan microcarriers for tissue engineering cartilage, Eur. Polym. J. 44 (2008) 2820–2829. https://doi.org/10.1016/j.eurpolymj.2008.06.021

[89]   K. Song, L. Li, W. Li, Y. Zhu, Z. Jiao, M. Lim, M. Fang, F. Shi, L. Wang, T. Liu, Three-dimensional dynamic fabrication of engineered cartilage based on chitosan/ gelatin hybrid hydrogel scaffold in a spinner flask with a special designed steel frame, Mater. Sci. Eng. C. 55 (2015) 384–392. https://doi.org/10.1016/j.msec.2015.05.062

[90]   A. Lahiji, A. Sohrabi, D.S. Hungerford, C.G. Frondoza, Chitosan supports the expression of extracellular matrix proteins in human osteoblasts and chondrocytes, J. Biomed. Mater. Res. 51 (2000) 586–595. https://doi.org/10.1002/1097-4636(20000915)51:4<586::AID-JBM6>3.0.CO;2-S

[91]   J.K. Francis Suh, H.W.T. Matthew, Application of chitosan-based polysaccharide biomaterials in cartilage tissue engineering: A review, Biomater. 21 (2000) 2589–2598. https://doi.org/10.1016/S0142-9612(00)00126-5

[92]   M. Sittinger, D.W. Hutmacher, M.V. Risbud, Current strategies for cell delivery in cartilage and bone regeneration, Curr. Opin. Biotechnol. 15 (2004) 411–418. https://doi.org/10.1016/j.copbio.2004.08.010

[93]   A. Di Martino, M. Sittinger, M.V. Risbud, Chitosan: A versatile biopolymer for orthopaedic tissue-engineering, Biomater. 26 (2005) 5983–5990. https://doi.org/10.1016/j.biomaterials.2005.03.016

[94]   X. Liang, X. Wang, Q. Xu, Y. Lu, Y. Zhang, H. Xia, A. Lu, L. Zhang, Rubbery chitosan/carrageenan hydrogels constructed through an electroneutrality system and their potential application as cartilage scaffolds, Biomacromol. 19 (2018) 340–352. https://doi.org/10.1021/acs.biomac.7b01456

[95]   S.H. Hsu, S.W. Whu, S.C. Hsieh, C.L. Tsai, D.C. Chen, T.S. Tan, Evaluation of chitosan–alginate–hyaluronate complexes modified by an RGD-containing protein as tissue-engineering scaffolds for cartilage regeneration, Artif. Organs. 28 (2004) 693–703. https://doi.org/10.1111/j.1525-1594.2004.00046.x

[96]   S. Chameettachal, S. Murab, R. Vaid, S. Midha, S. Ghosh, Effect of visco-elastic silk-chitosan microcomposite scaffolds on matrix deposition and biomechanical functionality for cartilage tissue engineering, J. Tissue Eng. Regen. Med. 11 (2017) 1212-1229. https://doi.org/10.1002/term.2024

[97]   R.S. Tigli, M. Gumusderelioglu, Evaluation of alginatechitosan semi IPNs as cartilage scaffolds, J. Mater. Sci. Mater. Med. 20 (2009) 699–709. https://doi.org/10.1007/s10856-008-3624-x

[98]   T. Freier, R. Montenegro, H. Shan Koh, M.S. Shoichet, Chitin-based tubes for tissue engineering in the nervous system, Biomater. 26 (2005) 4624–4632. https://doi.org/10.1016/j.biomaterials.2004.11.040

[99]   C.A. Heath, G.E. Rutkowski, The development of bioartificial nerve grafts for peripheral-nerve regeneration, Trends Biotechnol. 16 (1998) 163–168. https://doi.org/10.1016/S0167-7799(97)01165-7

[100] G. Ciardelli, V. Chiono, Materials for peripheral nerve regeneration, Macromol. Biosci. 6 (2006) 13–26. https://doi.org/10.1002/mabi.200500151

[101] M. Dash, F. Chiellini, R.M. Ottenbrite, E. Chiellini, Chitosan-A versatile semi-synthetic polymer in biomedical applications, Progress in Polym. Sci. 36 (2011) 981–1014. https://doi.org/10.1016/j.progpolymsci.2011.02.001

[102] C. Mingyu, G. Kai, L. Jiamou, G. Yandao, Z. Nanming, Z. Xiufang, Surface modification and characterization of chitosan film blended with poly-L-lysine, J. Biomater. Appl. 19 (2004) 59–75. https://doi.org/10.1177/0885328204043450

[103] Z. Zheng, Y. Wei, G. Wang, A.W.Q. Ao, Y. Gong, X. Zhang, Surface properties of chitosan films modified with polycations and their effects on the behavior of PC12 cells, J. Bioact. Compat. Polym. 24 (2009) 63–82. https://doi.org/10.1177/0883911508099653

[104] M. Cheng, J. Deng, F. Yang, Y. Gong, N. Zhao, X. Zhang, Study on physical properties and nerve cell affinity of composite films from chitosan and gelatin solutions, Biomater. 24 (2003) 2871–80. https://doi.org/10.1016/S0142-9612(03)00117-0

[105] A. Karimi, S. Karbasi, S. Razavi, E.N. Zargar, Poly(hydroxybutyrate)/chitosan aligned electrospun scaffold as a novel substrate for nerve tissue engineering, Adv. Biomed. Res. 7 (2018) 44. https://doi.org/10.4103/abr.abr_277_16

[106] H. Xu, Y. Yan, S. Li, PDLLA/chondroitin sulfate/chitosan/NGF conduits for peripheral nerve regeneration, Biomater. 32 (2011) 4506–4516. https://doi.org/10.1016/j.biomaterials.2011.02.023

[107] J.M. Gimble, F. Guilak, Differentiation potential of adipose derived adult stem (ADAS) cells, Curr. Top. Dev. Biol. 58 (2003) 137–160. https://doi.org/10.1016/S0070-2153(03)58005-X

[108] L. Aust, B. Devlin, S.J. Foster, Y.D. Halvorsen, K. Hicok, T. du Laney, A. Sen, G.D. Willingmyre, J.M. Gimble, Yield of human adipose-derived adult stem cells

from liposuction aspirates, Cytotherapy. 6 (2004) 7–14.
https://doi.org/10.1080/14653240310004539

[109] N.-C. Cheng, W.J. Lin, T.Y. Ling, T.H. Young, Sustained release of adipose-derived stem cells by thermosensitive chitosan/gelatin hydrogel for therapeutic angiogenesis, Acta Biomater. 51 (2017) 258–267.
https://doi.org/10.1016/j.actbio.2017.01.060

[110] Y.H. Cheng, S.H. Yang, F.H. Lin, Thermosensitive chitosan-gelatin-glycerolphosphate hydrogel as a controlled release system of ferulic acid for nucleus pulposus regeneration, Biomater. 32 (2011) 6953–6961.
https://doi.org/10.1016/j.biomaterials.2011.03.065

[111] N.C. Cheng, H.H. Chang, Y.K. Tu, T.H. Young, Efficient transfer of human adipose-derived stem cells by chitosan/gelatin blend films, J. Biomed. Mater. Res. B: Appl. Biomater. 100 (5) (2012) 1369–1377. https://doi.org/10.1002/jbm.b.32706

[112] A.M. Martins, G. Eng, S.G. Caridade, J.F. Mano, R.L. Reis, G. Vunjak-Novakovic, Electrically conductive chitosan/carbon scaffolds for cardiac tissue engineering. Biomacromol. 15 (2014) 635−643. https://doi.org/10.1021/bm401679q

[113] S. Saravanan, N. Sareen, E. Abu-El-Rub, H. Ashour, G.L. Sequiera, H.I. Ammar, V. Gopinath, A.A. Shamaa, S.S.E. Sayed, M. Moudgil, J. Vadivelu, S. Dhingra, Graphene oxide-gold nanosheets containing chitosan scaffold improves ventricular contractility and function after implantation into infarcted heart, Sci. Rep. 8(15069) (2018), 1-13. https://doi.org/10.1038/s41598-018-33144-0

[114] J. Li, J. Pan, L. Zhang, X. Guo, Y. Yu, Culture of primary rat hepatocytes within porous chitosan scaffolds, J. Biomed. Mater. Res. Part A 67A (3) (2003) 938–943.
https://doi.org/10.1002/jbm.a.10076

[115] X.H. Wang, D.P. Li, W.J. Wang, Q.L. Feng, F.Z. Cui, Y.X. Xu, X.H. Song, Mark van der Werf. Crosslinked collagen/chitosan matrix for artificial livers, Biomater. 24 (2003) 3213–3220. https://doi.org/10.1016/S0142-9612(03)00170-4

[116] F. Chen, M. Tian, D. Zhang, J. Wang, Q. Wang, X. Yu, X. Zhang, C. Wan, Preparation and characterization of oxidized alginate covalently cross-linked galactosylated chitosan scaffold for liver tissue engineering, Mater. Sci. Eng. C. 32(2) (2012) 310–320. https://doi.org/10.1016/j.msec.2011.10.034

[117] S.J. Seo, I.Y. Kim, Y.J. Choi, T. Akaike, C.S. Cho, Enhanced liver functions of hepatocytes cocultured with NIH 3T3 in the alginate/galactosylated chitosan scaffold, Biomater. 27 (2006) 1487–1495. https://doi.org/10.1016/j.biomaterials.2005.09.018

[118] Z. She, C. Jin, Z. Huang, B. Zhang, Q. Feng, Y. Xu, Silk fibroin/chitosan scaffold: Preparation, characterization, and culture with HepG2 cell, J. Mater. Sci. Mater. Med. 19 (12) (2008) 3545–3553. https://doi.org/10.1007/s10856-008-3526-y

[119] S. Zhending, L. Weiqiang, F. Qingling, Self-assembly model, hepatocytes attachment and inflammatory response for silk fibroin/chitosan scaffolds, Biomed. Mater. 4 (4) (2009) 045014. https://doi.org/10.1088/1748-6041/4/4/045014

[120] F.A. Shah, M. Trobos, P. Thomsen, A. Palmquist, Commercially pure titanium(cp-Ti) versus titanium alloy (Ti6Al4 V) materials as bone anchored implants-is one truly better than the other, Mater. Sci. Eng. C. 1 (2016) 960–6. https://doi.org/10.1016/j.msec.2016.01.032

[121] L. Zhang, K. Wu, W. Song, H. Xu, R. An, L. Zhao, B. Liu, Y. Zhang, Chitosan/siCkip-1 biofunctionalized titanium implant for improved osseointegration in the osteoporotic condition, Sci. Rep. 5 (2015) 10860. https://doi.org/10.1038/srep10860

[122] D.D. Divakar, N.T. Jastaniyah, H.G. Altamimi, Y.O. Alnakhli, A.A. Muzaheed, A.S. Haleem, Enhanced antimicrobial activity of naturally derived bioactive molecule chitosan conjugated silver nanoparticle against dental implant pathogens, Int. J. Biol. Macromol. 108 (2018) 790–797. https://doi.org/10.1016/j.ijbiomac.2017.10.166

[123] T. Sovány, A. Csüllög, E.Å. Benkå, G. Regdon Jr, K. Pintye-Hódi, Comparison of the properties of implantable matrices prepared from degradable and non-degradable polymers for bisphosphonate delivery, Int. J. Pharm. 533 (2017) 364–72. https://doi.org/10.1016/j.ijpharm.2017.07.023

[124] Z. Shariatinia, Pharmaceutical applications of chitosan, Adv. Colloid Inter. Sci. 263 (2019) 131–194. https://doi.org/10.1016/j.cis.2018.11.008

[125] K.C. Rani, R. Primaharinastiti, E. Hendradi, Preparation and evaluation of ciprofloxacin implants using bovine hydroxyapatite-chitosan composite and glutaraldehyde for osteomyelitis, Int. J. Pharm. Sci. 8 (2016) 45-51.

[126] S. Stefan, Combination of serum eye drops with hydrogels bandage contact lenses in the treatment of persistent epithelial defects, Graefes Arch. Clin. Exp. Ophthalmol. 244 (2006) 1345–9. https://doi.org/10.1007/s00417-006-0257-y

[127] S. Xin-Yuan, T. Tian-Wei, New contact lens based on chitosan/gelatin composites, J. Bioact. Compat. Polym. 19 (2004) 467-479. https://doi.org/10.1177/0883911504048410

[128] A. Bachhuka, J. Hayball, L.E. Smith, K. Vasilev, Effect of surface chemical functionalities on collagen deposition by primary human dermal fibroblasts, ACS Appl. Mater. Interfaces. 7 (2015) 23767–75. https://doi.org/10.1021/acsami.5b08249

[129] I. Kohsari, Z. Shariatinia, S.M. Pourmortazavi, Antibacterial electrospun chitosanpolyethylene oxide nanocomposite mats containing ZIF-8 nanoparticles, Int. J. Biol. Macromol. 91 (2016) 778–88. https://doi.org/10.1016/j.ijbiomac.2016.06.039

[130] Z. Shariatinia, A.M. Jalali, Chitosan-based hydrogels: Preparation, properties and applications, Int. J. Biol. Macromol. 115 (2018) 194–220. https://doi.org/10.1016/j.ijbiomac.2018.04.034

[131] I. Kohsari, Z. Shariatinia, S.M. Pourmortazavi, Antibacterial electrospun chitosan–polyethylene oxide nanocomposite mats containing bioactive silver nanoparticles, Carbohydr. Polym. 140 (2016) 287–298. https://doi.org/10.1016/j.carbpol.2015.12.075

[132] K. Bankoti, A.P. Rameshbabu, S. Datta, P.P. Maity, P. Goswami, P. Datta, S.K. Ghosh, A. Mitra, S. Dhara, Accelerated healing of full thickness dermal wounds by macroporous waterborne polyurethane-chitosan hydrogel scaffolds, Mater. Sci. Eng. C 81 (2017) 133–143. https://doi.org/10.1016/j.msec.2017.07.018

[133] P.T.S. Kumar, S. Abilash, K. Manzoor, S.V. Nair, H. Tamura, R. Jayakumar, Preparation and characterization of novel -chitin/nano silver composite scaffolds for wound dressing applications, Carbohydr. Polym. 80 (2010) 761–767. https://doi.org/10.1016/j.carbpol.2009.12.024

[134] K. Madhumathi, P.T.S. Kumar, S. Abhilash, V. Sreeja, H. Tamura, K. Manzoor, S.V. Nair, R. Jayakumar, Development of novel chitin/nanosilver composite scaffolds for wound dressing applications, J. Mater. Sci. Mater. Med. 21 (2010) 807–813. https://doi.org/10.1007/s10856-009-3877-z

[135] H.L. Lai, A. Abu'Khalil, D.Q.Craig, The preparation and characterisation of drugloaded alginate and chitosan sponges, Int. J. Pharm. 251 (2003) 175–181. https://doi.org/10.1016/S0378-5173(02)00590-2

[136] H.A. Hazzah, R.M. Farid, M.M.A. Nasra, M.A. EL-Massik, O.Y. Abdallah, Lyophilized sponges loaded with curcumin solid lipid nanoparticles for buccal delivery: Development and characterization, Int. J. Pharm. 492 (2015) 248–257. https://doi.org/10.1016/j.ijpharm.2015.06.022

[137] R. Jayakumar, M. Prabaharan, K.P.T. Sudheesh, S.V. Nair, H. Tamura, Biomaterials based on chitin and chitosan in wound dressing applications, Biotechnol. Adv. 29 (2011) 322–37. https://doi.org/10.1016/j.biotechadv.2011.01.005

[138] X. Wang, D. Zhang, J. Wang, R. Tang, B. Wei, Q. Jiang, Succinyl pullulan-crosslinked carboxymethyl chitosan sponges for potential wound dressing, Int. J. Polym. Mater. Polym. Biomater. 66 (2017) 61–70. https://doi.org/10.1080/00914037.2016.1182912

[139] S.G. Kumbar, K.S. Soppimath, T.M. Aminabhavi, Synthesis and characterization of polyacrylamide-grafted chitosan hydrogel microspheres for the controlled release of indomethacin, J. App. Polym. Sci. 87 (2003) 1525–1536. https://doi.org/10.1002/app.11552

[140] D. Altiok, E. Altiok, F. Tihminlioglu, Physical, antibacterial and antioxidant properties of chitosan films incorporated with thyme oil for potential wound healing applications, J. Mater. Sci. Mater. Med. 21 (2010) 2227–2236. https://doi.org/10.1007/s10856-010-4065-x

[141] S. Gaysinsky, P. Davidson, D. McClements, J. Weiss, Formulation and characterization of phyto-phenol-carrying antimicrobial micro emulsions, Food Bio. Physics. 3(1) (2008) 54–65. https://doi.org/10.1007/s11483-007-9048-1

[142] X. Li, S. Chen, B. Zhang, M. Li, K. Diao, Z. Zhang, J, Li, Y, Xu, X. Wang, H. Chen, In situ injectable nano-composite hydrogel composed of curcumin, N, O-carboxymethyl chitosan and oxidized alginate for wound healing application, Int. J. Pharm. 437(1–2) (2012) 110–119. https://doi.org/10.1016/j.ijpharm.2012.08.001

[143] D. Zhang, W. Zhou, B. Wei, X. Wang, R. Tanga J. Nie, J. Wang, Carboxyl-modified poly(vinyl alcohol)-crosslinked chitosan hydrogel films for potential wound dressing, Carbohydr. Polym. 125 (2015) 189–199. https://doi.org/10.1016/j.carbpol.2015.02.034

[144] M.J. Galotto, C.L. de Dicastillo, A. Torres, A. Guarda, Thymol: Use in antimicrobial packaging, in: J. Barros-Velázquez (Ed.), Antimicrobial Food Packaging, Academic Press, Cambridge, US, 2016, pp 553–562. https://doi.org/10.1016/B978-0-12-800723-5.00045-0

[145] R. Parhi, T. Panchamukhi, RSM based design and optimization of transdermal film of Ondansetron HCl, J. Pharm. Innov. 2019. https://doi.org/10.1007/s12247-019-09373-9. https://doi.org/10.1007/s12247-019-09373-9

[146] R. Parhi, P. Suresh, Transdermal delivery of Diltiazem HCl from matrix film: Effect of penetration enhancers and study of antihypertensive activity in rabbit model, J. Adv. Res. 7 (2016), 539–550. https://doi.org/10.1016/j.jare.2015.09.001

[147] S.K. Mishra, D.S. Mary, S. Kannan, Copper incorporated microporous chitosan-polyethylene glycol hydrogels loaded with naproxen for effective drug release and anti-infection wound dressing. Int. J. Biol. Macromol. 95 (2017) 928–937. https://doi.org/10.1016/j.ijbiomac.2016.10.080

[148] D. Archana, B.K. Singh, J. Dutta, P.K. Dutta, In vivo evaluation of chitosan–PVP–titanium dioxide nanocomposite as wound dressing material, Carbohydr. Polym. 95 (2013) 530–539. https://doi.org/10.1016/j.carbpol.2013.03.034

[149] P. Agrawal, G.J. Strijkers, K. Nicolay, Chitosan-based systems for molecular imaging, Adv. Drug Deliv. Rev. 62 (2010) 42–58. https://doi.org/10.1016/j.addr.2009.09.007

[150] C.M. Lee, H.J. Jeong, S.L. Kim, E.M. Kim, D.W. Kim, S.T. Lim, K.Y. Jang, Y.Y. Jeong, J.W. Nah, M.H. Sohn, SPION-loaded chitosan-linoleic acid nanoparticles to target hepatocytes, Int. J. Pharm. 371 (2009) 163–9. https://doi.org/10.1016/j.ijpharm.2008.12.021

[151] Z. Shi, K.G. Neoh, E.T. Kang, B. Shuter, S.C. Wang, C. Poh, W. Wang, (Carboxymethyl) chitosan-modified superparamagnetic iron oxide nanoparticles for magnetic resonance imaging of stem cells, ACS Appl. Mater. Interfaces. 1 (2009) 328–335. https://doi.org/10.1021/am8000538

[152] M. Koping-Hoggard, I. Tubulekas, H. Guan, K. Edwards, M. Nilsson, K.M. Varum, P. Artursson, Chitosan as a nonviral gene delivery system: Structure-property relationships and characteristics compared with polyethylenimine in vitro and after lung administration in vivo, Gene Ther. 8 (2001) 1108–21. https://doi.org/10.1038/sj.gt.3301492

[153] F.L. Mi, S.S. Shyu, C.K. Peng, Y.B. Wu, H.W. Sung, P.S. Wang, C.C. Huang, Fabrication of chondroitin sulfate-chitosan composite artificial extracellular matrix for stabilization of fibroblast growth factor, J. Biomed. Mater. Res. A 76 (2006) 1–15. https://doi.org/10.1002/jbm.a.30298

[154] J.H. Kim, Y.S. Kim, S. Kim, J.H. Park, K. Kim, K. Choi, H. Chung, S.Y. Jeong, R.W. Park, I.S. Kim, Hydrophobically modified glycol chitosan nanoparticles as carriers for paclitaxel, J. Contr. Release. 111 (2006) 228–234. https://doi.org/10.1016/j.jconrel.2005.12.013

[155]   S. Naskar, S. Sharma, K. Kuotsu, Chitosan-based nanoparticles: An overview of biomedical applications and its preparation, J. Drug Deliv. Sci. Technol. 49 (2019) 66–81. https://doi.org/10.1016/j.jddst.2018.10.022

[156]   J.H. Kim, Y.S. Kim, K. Park, S. Lee, H.Y. Nam, K.H. Min, H.G. Jo, J.H. Park, K. Choi, S.Y. Jeong, Antitumor efficacy of cisplatin-loaded glycol chitosan nanoparticles in tumor-bearing mice, J. Contr. Release. 127 (2008) 41–49. https://doi.org/10.1016/j.jconrel.2007.12.014

[157]   S. Mitra, U. Gaur, P.C. Ghosh, A.N. Maitra, Tumour targeted delivery of encapsulated dextran-doxorubicin conjugate using chitosan nanoparticles as carrier, J. Contr. Release. 74 (2001) 317–323. https://doi.org/10.1016/S0168-3659(01)00342-X

[158]   Y.J. Son, J.S. Jang, Y.W. Cho, H.C. Chung, R.W. Park, I.C. Kwon, I.S. Kim, J.Y. Park, S.B. Seo, C.R. Park, S.Y. Jeong, Biodistribution and anti-tumor efficacy of doxorubicin loaded glycol-chitosan nanoaggregates by EPR effect, J. Contr. Release. 91 (2003) 135–145. https://doi.org/10.1016/S0168-3659(03)00231-1

[159]   S. Parveen, S.K. Sahoo, Evaluation of cytotoxicity and mechanism of apoptosis of doxorubicin using folate-decorated chitosan nanoparticles for targeted delivery to retinoblastoma, Canc. Nano. 1 (2010) 47–62. https://doi.org/10.1007/s12645-010-0006-0

[160]   U. Termsarasab, H.-J. Cho, D.H. Kim, S. Chong, S.J. Chung, C.K. Shim, H.T. Moon, D.D. Kim, Chitosan oligosaccharide–arachidic acid-based nanoparticles for anti-cancer drug delivery, Int. J. Pharm. 441 (2013) 373–38. https://doi.org/10.1016/j.ijpharm.2012.11.018

[161]   P.J.P. Espitia, W.X. Du, R.D.J. Avena-Bustillos, N.D.F.F. Soares, T.H. Mchugh, Edible films from pectin: Physical-mechanical and antimicrobial properties - a review, Food Hydrocoll. 35 (2014) 287–296. https://doi.org/10.1016/j.foodhyd.2013.06.005

[162]   M. Aider, Chitosan application for active bio-based films production and potential in the food industry: Review, LWT-Food Sci. Technol. 43 (2010) 837-842. https://doi.org/10.1016/j.lwt.2010.01.021

[163]   G. Kerch, Chitosan films and coatings prevent losses of fresh fruit nutritional quality: A review, Trends in Food Sci. Technol. 46 (2015) 159-166. https://doi.org/10.1016/j.tifs.2015.10.010

[164] F. Shahidi, P. Ambigaipalan, Phenolics and polyphenolics in foods, beverages and spices: Antioxidant activity and health effects: A review, J. Func. Food. 18 (2015) 820-897. https://doi.org/10.1016/j.jff.2015.06.018

[165] J. Garrido, F. Borges, Wine and grape polyphenols-A chemical perspective, Food Res. Int. 54 (2013) 1844–1858. https://doi.org/10.1016/j.foodres.2013.08.002

[166] D. Kadam, S.S. Lele, Cross-linking effect of polyphenolic extracts of Lepidium sativum seedcake on physicochemical properties of chitosan films. Int. J. Biol. Macromol. 114 (2018) 1240–1247. https://doi.org/10.1016/j.ijbiomac.2018.04.018

[167] H. Yong, X. Wang, R. Bai, Z. Miao, X. Zhang, J. Liu, Development of antioxidant and intelligent pH-sensing packaging films by incorporating purple-fleshed sweet potato extract into chitosan matrix, Food Hydrocoll. 90 (2019) 216–224. https://doi.org/10.1016/j.foodhyd.2018.12.015

[168] J. Liu, S. Liu, Q. Wu, Y. Gu, J. Kan, C. Jin, Effect of protocatechuic acid incorporation on the physical, mechanical, structural and antioxidant properties of chitosan film, Food Hydrocoll. 73 (2017) 90-100. https://doi.org/10.1016/j.foodhyd.2017.06.035

[169] J. Liu, C.g. Meng, S. Liu, J. Kan, C.h. Jin, Preparation and characterization of protocatechuic acid grafted chitosan films with antioxidant activity, Food Hydrocoll. 63 (2017) 457-466. https://doi.org/10.1016/j.foodhyd.2016.09.035

[170] K. Rambabu, G. Bharath, F. Banat, P.L. Show, H.H. Cocoletzi, Mango leaf extract incorporated chitosan antioxidant film for active food packaging, Int. J. Biol. Macromol. 126 (2019) 1234–1243. https://doi.org/10.1016/j.ijbiomac.2018.12.196

[171] A. Riaz, S. Lei, H.Md.S. Akhtar, P. Wan, D. Chen, S. Jabbar, Md. Abid, M.Md. Hashim, X. Zeng, Preparation and characterization of chitosan-based antimicrobial active food packaging film incorporated with apple peel polyphenols, Int. J. Biol. Macromol. 114 (2018) 547–555. https://doi.org/10.1016/j.ijbiomac.2018.03.126

[172] E. Talón, K.T. Trifkovic, M.Vargas, A. Chiralt, C. González-Martínez, Release of polyphenols from starch-chitosan based films containing thyme extract, Carbohydr. Polym. 175 (2017) 122–130. https://doi.org/10.1016/j.carbpol.2017.07.067

Advanced Applications of Polysaccharides and their Composites
Materials Research Foundations **73** (2020) 136-183

Materials Research Forum LLC
https://doi.org/10.21741/9781644900772-5

Chapter 5

Nanocellulose-Improved Food Packaging

Elaine Cristina Lengowski[1], Eraldo Antonio Bonfatti Júnior[2], Kestur Gundappa Satyanarayana[3*]

[1]College of Forestry Engineering. Federal University of Mato Grosso (UFMT) Fernando Corrêa da Costa St, 2367 - Boa Esperança Cuiabá MT 78068-600 Brazil

[2]Department of Forest Engineering and Technology. Federal University of Paraná (UFPR) Pref. Lothário Meissner Avenue, 632, Jardim Botânico Curitiba PR 80210-170 Brazil

[3]Poornaprajna Scientific Research Institute (PPISR) Sy. no. 167, Poornaprajnapura Bidalur Post Devanahalli, Bangalore 562 110 Karnataka India

* gundsat42@hotmail.com, kgs_satya@yahoo.co.in

Abstract

Although traditional food packaging has contributed to the early development of the distribution system, this is not sufficient to meet new consumer requirements. The application of nanocellulose in packaging and coatings promises to open up new possibilities to improve not only the properties of the packaging, but also the efficiency, contributing to the preservation of fresh food by extending its shelf life and reducing environmental waste. The main objective of this chapter is to present the scientific information of relevance regarding nanocellulose and the development of composites applied to the food packaging sector.

Keywords

Nanotechnology, Natural Fibers, Smart Packaging

Contents

1.   Introduction

All humans need to ingest nutrients to survive. This ingestion is made through substances whether processed, semi-processed or raw, including beverages, chewing gum and any other products that are used in the manufacture or preparation known as food [1]. Foods can be of animal or vegetable origin, and both contain nutrients essential to humans, being carbohydrates, fats, proteins, vitamins, etc. [2]. However, although there are only two sources, other types of food can be prepared, and these are called processed food and functional food [3,4].

Linked to the increase in world population is the increase in demand for food, which certainly requires some form of preserving this food. For sustainable food system, maintenance of its quality and improved safety are the important factors whereby postharvest losses can be reduced [5]. If these are not taken care of, there would be high losses in postharvest leading to large amount of wastage of food, which would become a great problem to the food industry all over the world. For example, it is reported that annual wastage of about 1.3 billion tons of food in production, distribution and at homes [6], while about 33% purchased food by some organizations and houses in some of the developed countries such as Britain, Sweden and USA is wasted [7]. It is also reported that such loss and wastage of food would lead to insecurity of food all over the world besides a wide range of socio-economic, climatic and environmental factors [5]. There are also environmental benefits by the reduction of losses and wastage of food, which is underlined by the fact that emission of about 4.2 tons of CO_2 per one tons of waste food [6].

To take care of the above mentioned losses and wastage of food even in transport for short distances or to keep the food for later use, even in olden days people used to take care of the safety of food using either leaves from plants, containers made of wood or mud and even animal skin as packaging materials [8]. With the progress of time, technology of processing of food and its packaging has gone through significant changes. In the case of packaging, it has gone from simple and important function of containers and protection from 3 major external factors (physical, chemical and biological) to take care of a key marketing role through development of shelf appeal, providing product information, and establishing brand image and awareness besides conferring convenience to handle the food [9–14], whereby it can be consumed within the useful life of the food [15]. By the functions of preserver and protector, packaging helps in increasing the shelf life of the food, whereby it plays an important role in reducing wastage of food from the field or location of its preparation till its use [16].

It can be said that the packaging acts also as an information medium for producers of the food and its distributors through printing on it about name of the product, identification of its producers and date of manufacture and storage along with expiration date for its consumption, quantity of its content, nutritional information and home storage, preparation and instructions for its use, etc. [9,10]. Other additional information normally given on the packages include ease of opening the package, possibility of heating/cooking and serving of the food in the package itself, possibility of use in microwave ovens to heat the food before consumption and its suitability for different occasions of consumption [9].

In addition, the beneficial effects of processing, extension of shelf-life, and maintenance or increasing the quality and safety of food are attractive factors. All the above in turn has led to use of new packaging materials, new methods of manufacturing of these packaging materials, etc. It may be said that enhanced interest of development of packaging technology has been due to new demands on packaging, which in turn are due to newer innovations in food processing. It is reported that use of food packaging has led to increased spending by all strata of the society thus becoming a socioeconomic indicator of the population [17].

It is well known that most marketed foods are packed suggesting that packaging plays a key role in the food industry because of its multiple functions [9–11] as container, preserver, protector, informer besides conferring convenience to handle the food [12,13]. By the functions of preserver and protector, packaging helps in increasing the shelf life of the food, whereby it plays important role in reducing wastage of food from the field or location of its preparation till its use [16]. These developments have followed recent advances in biotechnology, nanotechnology and material science, which open up new possibilities for developing food packaging materials by the involvement of researchers in material science, food chemistry and technology and microbiology in the fields of food packaging and shelf-life.

From the foregoing it becomes evident that through various functions of packaging which include the safety of both fresh and processed food to the whole range of food supply chain till it reaches the final user, packaging plays an important role in both the food industry as well as in modern consumer marketing.

It is well known now that nanotechnology is one branch of technology, which deals with dimensions and tolerances lying between 1 to 100 nanometers (nm). Nanomaterial is one of the recent developments in the area of materials. These are produced from various resources of which biomass-based resources seem to be more attractive in view of abundant availability, low cost and environmentally friendly. Of these, biomass-based materials, cellulose is more fascinating and thus an attractive material to produce nanomaterials, knowing that it is almost inexhaustible sustainable material having remarkable properties for possible applications in several materials. Nano materials produced based on natural cellulose (these are from lignocellulosic biomass, bacteria and invertebrate sea creatures) are called 'nanocellulose', which commonly refers to three types of nano materials, viz, cellulose nanocrystals (CNCs), nanofibrils of cellulose (NFCs) and bacteria nanocellulose (BNC). These nanocelluloses exhibit exceptional and excellent physical, mechanical, electrical, and biological properties, besides being biocompatible, biodegradable, and possessing low cytotoxicity. In view of these, it is no wonder that these nanomaterials are more sought-after candidates in many potential applications in many areas of science and technology particularly as sustainable materials in the gamut of various industries. Food industry has not lagged behind in this new technology and use of nanotechnology in this industry has been projected to increase multifold in the coming years.

Advanced Applications of Polysaccharides and their Composites   Materials Research Forum LLC
Materials Research Foundations **73** (2020) 136-183   https://doi.org/10.21741/9781644900772-5

Considering the above-mentioned points on the food industry particularly in respect of packaging and use of nanotechnology in this, this Chapter presents various types or categories of packaging, the materials used for the production of these packaging with emphasis on nanocellulose in food packaging presenting the benefits of using the most different types of nanocelluloses in this segment.

2. Categories of packaging

There are three types of packaging, viz., primary, secondary and tertiary. Primary packaging is the term used to designate the packaging layer in direct contact with the product, normally responsible for the storage and preservation of this product [9,18,19], such as bottles and beverage cans. When used in food, these packings should be made from inert and non-toxic materials and therefore should not cause any changes in food, such as changes in color and taste [18].

Secondary packaging is that when the packaging contains one or more primary packaging [19], which is usually responsible for the physical-mechanical protection of food during its distribution [9,18]. Secondary packaging is also used to prevent the primary from getting dirty or contaminated [18] and is often responsible for the communication, being the information support, especially in cases where it contains only a primary packaging [9]. An example of secondary packaging is the plastic rings holding a set of cans or the carton containing cans.

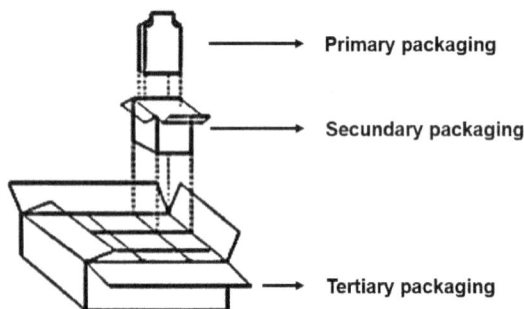

Figure 1. Graphical representation of the packaging systemwith primary, secondary and tertiary packaging.

Finally, the tertiary packaging is the one that groups many primary or secondary packaging for transport [9,19], also called "distribution package", to protect the product during distribution and to help in efficient handling [18]. A corrugated box is the most common form of tertiary package [9,18]. A graphical representation of packaging system with primary, secondary and tertiary packaging is shown in Fig. 1.

A group of tertiary packages mounted on a single base is called 'unit load', which and aims to facilitate the automation and transport of large loads [18], a pallet jack or a ship containers can be considered an unit load.

3. Types of food packaging

Looking at the development of packaging over the years, it might have started with wrapping of food, which probably did not have scientific basis, except for the consumer in the eyes of the sellers. For this purpose, either paper or cloth in the form of sacks might have been used in olden days. As the time progressed and with the development of science, man started using different types of packaging, which include glass, plastics (cellophane and polyethylene), metals (for beverages), etc. Reason for this could be that each food has specific needs in terms of its storage, storage and transport. Therefore, it is necessary to know different types of the packaging in order to choose the ideal model for each type of food. Various types of available packaging are discussed in the following subsections.

3.1 Conventional food packaging

Conventional packaging are those which are limited to the basic functions of containing, preserving, protecting, informing and conferring convenience [9,12], with minimal interaction between the packaging and the food during the storage and distribution stages. Although conventional packaging has greatly contributed to the early development of food distribution systems, it is no longer efficient because it does not respond to consumer demands for minimally processed foods containing low quantity of preservatives [12,20].

3.2 Active food packaging

Food producers are generally challenged to produce high quality food having longer shelf life in comparison to ordinary food. Such food should therefore be packed using suitable packaging material. One such method is 'active' packaging.

Active packaging systems were defined by Rooney [21] as a material that plays a different role from an inert barrier to the external environment. Therefore active packaging has several additional functions over conventional packaging by changing product conditions, increasing shelf life, safety, quality and/or improving its sensory characteristics [22].

Active packaging allows the interaction between the packaging, the food and the internal and external environments to improve sensory properties and food safety, while ensuring maintenance of food quality [23–25].

In recent years, research to develop active packaging production techniques has significantly increased [26], the focus of these research is to find additives that can be incorporated into packaging instead of direct incorporation into food [27]. These additives absorb undesirable substances such as oxygen, ethylene, moisture, light, dust, odor, heavy metals, depleted oils, free radicals and other contaminants [20,28–30], and some additives emit carbon dioxide, antimicrobial agents, antioxidants and aromatics [23,28,31].

Examples of active packaging are humidity regulators, ethylene absorbers, films containing antimicrobial and antioxidant agents, enzymes, flavorings, packaging systems, or emitters of flavoring agents, systems that allow monitoring of product quality beyond temperature and other characteristics physicists [32–35].

There are two aspects in active packaging, viz., active scavenging and active releasing packaging system [5]. In the first one unwanted materials (oxygen, excessive moisture, etc.) are removed while in the second beneficial agents such as carbon dioxide, moisture, preservatives and antioxidants added with a view to preserve the quality of its content. Ensuring exceptional food quality and extended shelf-life are the main objective of both these systems.

3.2.1 Antimicrobial food packaging

Foods can serve as a vehicle and substrate for the multiplication of several microorganisms, often pathogenic, capable of producing toxins and causing health risk to the consumer when ingested [36], and the presence of deteriorating microorganisms in food is one of the main reasons for the loss of food [37]. In order to avoid biological contamination of food, the concept of antimicrobial packaging is presented, which presents substances incorporated in the packaging material capable of eliminating or inhibiting deteriorating and pathogenic microorganisms of food through slow release to the food surface or it may be used in vapor form [8-12]. It is reported that some of the materials such as silver nanoparticles, ethyl alcohol, Chlorine dioxide, Nisin, organic acids, etc. are being evaluated for the antimicrobial properties [17].

Thus, the antimicrobial package is an active packaging that acts to reduce, inhibit or retard the growth of microorganisms that may be present in the packaged food itself or in the packaging material itself [36]. The basic principle of antimicrobial packaging is the addition of an extra microbiological barrier to the physical barriers of conventional packaging [38].

According to Appendini and Hotchkiss [36] the strategies for antimicrobial packaging are:

1. Addition of sachets containing volatile antimicrobial substances into the packages.

2. Incorporation of volatile and non-volatile antimicrobial substances directly into polymers.

3. Coat or adsorb antimicrobial substances on polymer surfaces.

4. Immobilization of antimicrobial substances in polymers by ionic or covalent bonds.

5. Use of polymers that are inherently antimicrobial.

From the above it can be said use of antimicrobials in food packaging would result in enhancing the quality and safety of the food through the reduction of surface contamination of the food [17].

3.2.2 Antioxidant food packaging

Oxidation is the most frequent mechanism of deterioration and reduced shelf life [28], especially for lipid-rich foods [39–43] such as meats and their derivatives [44–46], milk and dairy products [47]. In addition to altering the taste, oxidation decreases the nutritional quality of the food [44–46], since vitamins and fatty acids are degraded, and this degradation may have reactive and toxic compounds that can put consumers' health at risk [48].

In order to avoid or reduce food oxidation, both the food industry and the scientific community have been working on the development of new packaging systems to prolong the shelf life of food [49]. The most promising solution to preserver oxidation-sensitive foods are the active packaging antioxidants, or simply packaging antioxidants [32].

This type of active packaging has the strategy of adding antioxidant substances in which they interact with packaged food, limiting or preventing their [50,51], avoiding the addition of chemical compounds directly in foods [52]. To this end, the antioxidant compounds must be added in films, papers or sachets from where they will be released to protect foods from oxidative degradation, inhibiting oxidation reactions by reacting with free radicals and peroxides and consequently extending shelf life [22,28,53].

3.2.3 Aromatic food packaging

The incorporation of flavorings in food packaging improves the organoleptic quality of the product [22,29], improving sensory acceptance of conditioned food [54] through the controlled release of desirable flavors from polymeric materials [55]. The production strategy of these packages is the incorporation of volatile aromatic compounds into the polymer matrix [54]. The volatility of an aromatic packaging is fundamental for its effectiveness, such characteristic being dependent on the molecular weight of aromatic compounds, which, in most cases, has a chemical structure of 6 to 18 carbons [55].

The development of packaging that can contribute to the improvement of flavor through the release of flavoring compounds to food is a marketing alternative, since the sensorial quality of food is, in most cases, the primary factor in the acquisition of a particular product. Therefore, flavor packaging is a good alternative for the development of new products or for the introduction of those already traditional with different flavors, allowing, therefore, a greater diversification of products in an increasingly competitive and demanding consumer market [56].

3.2.4 Smart/intelligent food packaging

Functions such as traceability, indication of breach and monitoring of food quality are increasingly gaining importance in the production of food packaging [32,57], these new systems are called intelligent packaging [58]. Smart packaging may have its function activated and deactivated in response to changes in external and internal conditions and may include communication to customers or end users regarding product status [59], briefly a smart packaging reads and reports the conditions of the feed [58].

Intelligent packaging can be divided into two groups: data carrier packages, where the bar code and radio frequency identification (RFID) labels are inserted, and indicator packages, with a special emphasis on the time-temperature binomial indicators, indicators of gases such as oxygen and ethylene, and indicators of pathogenic microorganisms and toxins [12,60].

Figure 2 shows some bar codes being used. The most popular bar code (Fig. 2a) was created in 1970 [61], also known as "Universal Product Code" (UPC) is still widely used in the world due to its low cost [62]. However, new bar code models, such as QR code (Fig. 2b) and data matrix (Fig. 2c), with greater capacity to store data, have arisen [12,62], since these models store data both directions, vertical as well as horizontal [62].

On the other hand, Fig. 3 shows a new technology for data storage, automatic product identification and traceability, which is called 'Radio Frequency Identification (RFID) tags. This system basically consists of a microchip connected to a thin antenna [63–65].

Regarding the bar code, the RFID tag has some advantages because it does not require direct contact with the reader and has a greater capacity to store data, which facilitates its association with indicators and biosensors [58,66,67].

(a)   *(b)*   *(c)*

Figure 2. Three popular barcode patterns: (a) Universal Product Code (UPC), (b) quick Response code (QR code), (c) data matri. (Reproduced from Chen et al.[61] with the kind permission of the Publishers).

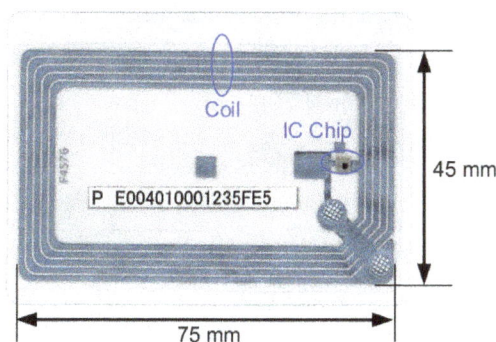

Figure 3. Example of a RFID tag. (Reproduced from Fujisaki [63] with the kind permission of the Publishers).

The uses of RFID in the food industry are numerous and range from facilitating food traceability to improving the efficiency of supply chains, but perhaps the ultimate benefits of RFID in food packaging are that it speeds up inventory turnover and improves traceability [58], in addition to the possibility of its association with biosensors [68–70].

It may be noted that the 'indicators' are compounds that indicate the presence, absence or concentration of another substance or the degree of reaction between two or more substances [71]. According to Biji et al. [72] in the food industry can be highlighted three types of indicators:

1. Freshness indicators: provide information on the quality of the product in relation to microbial growth or chemical changes within the food product.

2. Time temperature indicators (TTI): These indicators indicate whether the temperature limit has been exceeded over time and the time that a product has passed the temperature limit.

3. Integrity indicators: it has the function of informing leaks or ruptures in the packaging, guaranteeing the quality of the food throughout the chain of production and distribution.

Table 1 shows some indicator systems in smart packaging available on the market.

Table 1. Commercially available indicators food packaging systems.(adapted from Biji et al. [71] with the kind permission of the Publishers).

Trade Name	Manufacturer	Indicator Type
O$_2$ Sense™	Freshpoint Lab	Integrity
Novas®	Insignia Technologies Ltd.	Integrity
Ageless Eye®	Mitsubishi Gas Chemical Inc.	Integrity
Freshtag®	COX Technologies	Freshness
Sensorq®	DSM NV and Food Quality Sensor Int.	Freshness
Timestrip Complete®	Timestrip UK Ltd.	Time temperature
Timestrip® PLUS Duo	Timestrip UK Ltd.	Temperature
Monitormark™	3M™	Time temperature
Fresh-Check®	Temptime Corp.	Time temperature
Onvu™	Ciba Specialty Chemicals AndFreshpoint	Time temperature
Checkpoint®	Vitsab	Time temperature
Cook-Chex	Pymah Corp.	Time temperature
Colour-Therm	ColourTherm	Time temperature
Thermax	Thermographic Measurements Ltd.	Time temperature
Timestrip®	Timestrip Ltd.	Integrity

3.3   Edible films and coatings

'Films' and 'coatings' are terms used in the field of food, most often without distinction. However, it is important to note the difference between these two terms: the protection is a film formed by drying the biopolymer solution prepared separately from the food, which is then applied; while the coating may be a suspension or an emulsion applied directly to the food surface which upon drying leads to the formation of a film [73,74].

Both the edible films (EF) and edible coatings (EC) have attracted increasing attention in the last decade as a new alternative to reduce synthetic packaging and improve biodegradability [75]. Applied on the surface of the food seeking the reduction in the loss of water vapor, oxygen, migration of lipids and aroma or to stabilize the gradients of water activity and thus maintain different texture properties [73,76], it is possible to incorporate active agents such as antioxidants and antimicrobials to prolong the useful life of the food [53,77].

Various materials have already been used for these purposes, in China wax was used as fruit coatings in the twelfth and thirteenth centuries [78], in the 1930s paraffin waxes were commercially gained and were used in coating apples and pears [79]. Researches with biopolymer-based coatings such as polysaccharides, proteins and lipid materials have increased [74,80], emphasizing starch and nanocellulose due to their low costs, availability and biodegradability [74,81].

4.   Materials for food packaging

In recent times packaging technologies have become innovative. Accordingly, some of such technologies include 'modified atmosphere packaging', 'active packaging', and 'smart and intelligent packaging'. Extended shelf-lives, reduction in the rate of quality loss and increased safety of packaged foods are the results from these new technologies. Therefore, selection of appropriate packaging materials and systems is an integral part of food processing and product design [19]. Over the years, the food industry has always shown great interest in developing packaging with different materials, some toxic to humans and others not so [23]. In addition to the development of these new materials, the concepts of 'smart' packaging and 'active' packaging have been introduced [24,82], these have brought in new functions and applications to packaging [82].

The selection of the packaging system for a food depends on many factors such as the type of food, the protection requirements, the shelf life required for the food, the market for which it is intended, logistics, storage and sale [9,83]. The types of materials for food packaging are discussed in the following subsections.

4.1   Metals

Metal packaging was developed at the beginning of the 19th century, but it was from 1920 onwards that it started the application of metals in food packaging, using internal varnishes suitable for different types of products [9]. The metal packs offer the advantage of superior mechanical strength, impermeable barrier to mass and light transfer, good thermal conductivity and resistance to relatively high temperatures, these latter properties make the metal packages particularly suitable for thermal processing in the package [83,84], but metals are not suitable for the production of modified atmosphere packaging [85].

Four types of metals are commonly used in food packaging, viz., steel, aluminum, tin and chrome [18], but steel and aluminum are the most commonly used for this purpose [14], with food cans being mostly made of steel while beverage cans are generally produced from aluminum [18]. Unlike aluminum, steel tends to oxidize when exposed to moisture and oxygen, producing rust [18,84], and to avoid oxidation, tin and chromium are used as protective layers for steel [18].

4.2   Glass

Glass is an amorphous inorganic melt product which has been cooled to a rigid condition without crystallization [86], the production of glass containers involves heating a mixture of silica (the glass former), sodium carbonate (the melting agent) and limestone / calcium carbonate and alumina (stabilizers) at high temperatures until the materials merge into a thick liquid mass which is then poured into molds [14], in addition to these other compounds can be combined to obtain different types of glass [9,18].

As material for food packaging glass has interesting properties as optical transparency, excellent chemical stability, inert to many foods, very rigid and, because it is not permeable, has excellent barrier properties [14,18,84,87], to these advantages can be added the facts that foods can be cooked in glass receptors at high temperatures and that most glass containers can be reused and recycled [14,83,84].

However, there are some disadvantages of using glass as packaging material, which includes brittleness, because it breaks easily with high pressure, thermal shock and shock, and high weight [14,18,83,84]. For food packaging, fragility causes safety concerns because there is a possibility of chipped glass in food products [18], and high weight may increase transport and logistics costs [14].

Glass has long been used for food packaging, the first record of its use to hold food dates back to around 3000 BC [13,88], with a lower standardization degree than metal packaging [84] use of food packaging glass has declined in the last three decades, losing market share to metal cans and, increasingly, to plastics [18]. However, it is still an important packaging material [18], especially for beverages.

4.3 Paper

Paper is a two-dimensional material, which is produced from an aqueous suspension of fibers of wood origin, which in turn are artificially intertwined and subsequently dewatered by mechanical and thermal processes [89]. Paper production began more than two millennia, due to the need to communicate through materials that could be easily transported [90], however the paper application expanded, and today this material is used for other purposes, such as writing, printing, hygiene and packaging [81].

The first record of the use of paper as packaging is 1850 [91], after this milestone the packaging industry had for a long time paper as the main raw material [83], and is widely used for food packaging [84]. Today, although still mostly used, the paper divides this protagonism with other materials as the polymers [18].

The main advantages of using paper for food packaging are low cost, wide availability, low weight, printing capacity, good mechanical strength, ability to impart rigidity and flexibility when necessary, possibility of recycling and biodegradability [83,84,92]. As disadvantage, paper is not a material with good barrier properties, because its hydrophilic nature [92] and porous characteristics [91]. These render the paper permeable to moisture and gases. But this disadvantage can be circumvented by the association of paper with other materials, such as use of wax over the paper [84] and coating or incorporating nanocellulose (nanopaper) [81], and some strategies of paper waterproofing may hinder its recycling and biodegradability [92].

4.4 Synthetic polymers

Polymers are the most used materials for packaging production due to the low cost, the ease of process and the availability of abundant resources for its production [93]. Polymeric materials are quite varied and versatile, can be flexible or rigid, transparent or opaque, thermosetting or thermoplastic (thermosetting), reasonably crystalline or practically amorphous, can be produced as films or as containers of many shapes and sizes [84].

The most relevant polymer properties for packaging applications can be broadly classified into morphology, barrier properties, mechanical properties, thermal properties and optical properties [93]. Use of polymeric materials in food packaging requires low migration of chemical components from the packaging to the food and good barrier properties such as resistance to water, gas and liquid vapor permeability [9,93]. The common polymer used for food packaging application is shown in Fig. 4.

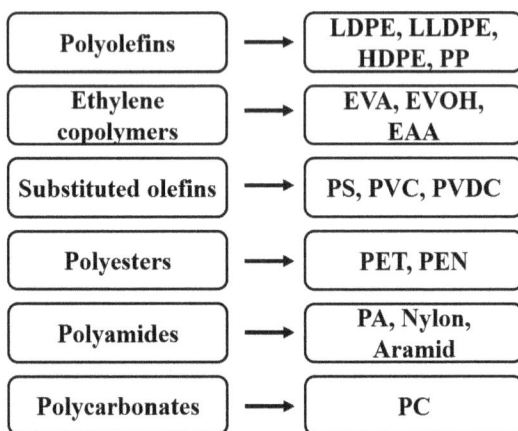

Figure 4. Common polymers used for food packaging application.LDPE: Low-density polyethlen; LLDPR: Linear low-density polyethylene; HDPE: High-density polyethylene; PP: Polypropylene; EVA: Ethylene-vinyl acetate; EVOH: Ethylene-vinyl alcohol; EAA: Ethylene acrylic acid copolymer; PS: Polystyrene; PVC: Polyvinyl chloride; PVDC: Polyvinylidene chloride; PET: Polyethylene terephthalate; PEN: Polyethylene naphthalate; PA: Polyamide; PC: Polycarbonate.

4.5   Biopolymers

Used food packaging constitutes a considerable and growing proportion of solid urban waste [84], polymers, metal, glass and some types of paper are recyclable. However, in the face of the large amount of waste produced, recycling is not enough and these materials become a source of contamination and pollution to the environment, especially the oceans, which are rapidly packing tons of plastic [94]. This leads to unwanted environmental issues, which are increasingly important to consumers and, as a consequence, there has been increase in research and production of bio-based and biodegradable materials.

Biologically based packaging is defined as packaging containing raw materials from agricultural and renewable sources [95], such as those produced from starch and cellulose [81]. Biopolymer based on cellulose and starch is not new, with cellophane being the most common cellulosic biopolymer and amylose, hydroxyl propylated starch and dextrin biopolymers based on starch [14]. Renewable polymers are classified according to their production method: polymers extracted directly from natural materials, polymers produced by "classical" chemical synthesis from renewable bio-derived monomers and polymers produced by micro-organisms or genetically modified bacteria [95].

Biopolymers are not as plastic, so it is not simple to substitute the production of plastic packaging for the production of biologically based packaging [14]. This is because, in addition the high cost of production, poor barrier properties, low mechanical resistance and low chemical compatibility with others polymers limit the use of these materials [14,95,96]. However, biopolymers increase the possibility of production of packaging and edible films [84] creating a positive outlook in a market dominated by materials from practically non-degradable fossil sources [96].

Bio-nanocomposites are being developed and used to reduce or mitigate the disadvantages of biopolymers and also improve the properties of traditional materials applied in the production of food packaging. Nanocellulose is a bio-nanocomposite that has been showing promising results, and its application in food packaging is discussed in the next section.

5.   Nanocellulose (NC)

The term "nanocellulose" began to be used in the 1980s, although the first studies in 1970 already pointed to a class of fibrillated materials, produced from cellulose pulp through micro processing, in which a microfibrils gel was obtained [97,98].

As mentioned earlier, nanocellulose is a general term that refers to a class of materials that has at least one of its dimensions on a nanometric scale [81,99]. Nanocelluloses can be produced by different sources and routes of production, such as fibrillation and hydrolysis (chemical or enzymatic) or their combinations [99–101].

Considering the top-down route, nanocelluloses can be classified into three types: nanocrystalline cellulose (CNC), microfibrillated cellulose (MFC) and nanofibrillatedcellulose (NFC) [101,102]. There is also bacterial nanocellulose (BNC)

[98,103], obtained by the bottom-top route, but its application will not be considered in this chapter.

MFC and NFC have crystalline and amorphous regions, differing in their size, MFC are fibers with a diameter between 25 and 100 nm, whereas NFC has diameters between 5 and 30 nm and a length between 2 and 10 μm [102,104], it is common to see the confusion between these two terms in the literature [100]. The CNC has no amorphous zones and has diameters between 2 and 20 nm and a length between 100 and 600 nm [102,105–107].

The CNC is produced by strong acid hydrolysis under controlled conditions of temperature, time and agitation followed by washing, centrifugation, dialysis and sonication to form a suspension followed by freeze-drying or heat drying. The electrostatic stability of the crystals depends on the type of acid used [106,108], and sulfuric acid and hydrochloric acid are commonly used [100]. Due to the presence of negative charges on the surface of the crystals, the CNCs produced with sulfuric acid are more stable and dispersed in water. On the other hand, CNCs produced with hydrochloric acid are more likely to clump together in aqueous solution [106]. Another useful feature of NC is its low coefficient of thermal expansion (CTE), which can be as low as 0.1 ppm K^{-1} and comparable to that of quartz glass [109].

Another method that can be employed, but which has less prominence in the CNC probe is enzymatic hydrolysis [110]. However, this method is expensive and time-consuming [111], and is now used as a pretreatment for the production of MFC or NFC.

NFC and MFC are produced by mechanical defibrillation of cellulose fibers. Some methods that may be employed in this fibrillation are grinding, cryocrushing with high-pressure homogenization with liquid nitrogen, steam explosion, high-intensity ultrasound etc. [81]. It may be noted that some pre-treatments may be required before mechanical processes for spending lower energy for producing NCs and smaller NCs with a view to promote the accessibility of hydroxyl groups, increase the internal surface, alter crystallinity, break the hydrogen bonds of the cellulose and thus increase the reactivity of the fibers. Among the pre-treatments that are used include enzymatic, chemical or alkaline hydrolysis [112].

Fig. 5 illustrates the hierarchical structure of the cellulose and the location of the structures that make up the MFC, NFC and CNC.

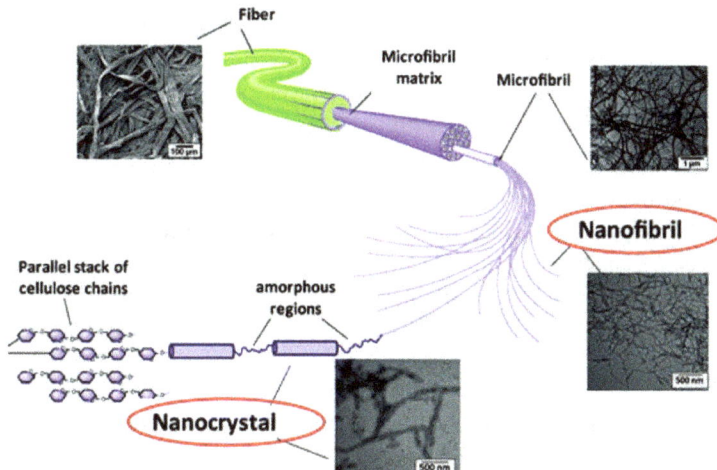

Figure 5. Hierarchical structure of cellulose. (Reproduced from Nelson et al. [105] with the kind permission of the publishers.

6. Use of nanotechnology in food packaging

In order to reduce the costs of petroleum-based polymers and to maintain or improve the properties of the packaging, research began in the 1960s on the development of composite materials with the addition of fibrous or particulate mineral fillers. Among the properties of the charge to be evaluated, this physical-chemical interaction, property directly tied to the particle size [113,114]. Due to this the application of nanomaterials in food packaging promises to open new possibilities to improve not only the properties but also the efficiency, contributing to the preservation of fresh foods by extending their shelf life. The nanometric size results in the marked improvement of the properties of flexibility, durability, biodegradability, temperature and humidity stability, and gas barrier properties when compared to conventional packages [96,115].

The nanotechnology applied in food packaging materials results in the following classification: (1) improved, whose nanomaterials can alter their properties and increase the commercial validity of the food; (2) active by the addition of nanomaterials, for example metal oxide or metal nanoparticles with antimicrobial properties; (3) bioactive packaging, with incorporating bioactive compounds capable of preventing or reducing disease risks and controlled release of substances; (4) intelligent, built-in nanosensors to monitor and report the conditions of the packaged food or surrounding environment [116]. In the context of improved packaging, the addition of nanocellulose has shown good gains in terms of mechanical, thermal and moisture barrier properties, as well as being biodegradable and low cost [117–120].

Nanocellulose is a promising material in food packaging applications due to its mechanical and barrier properties [119,120]. These properties are associated with the resistance of the crystalline regions and the strong network formed by the interfibrillar connections of the amorphous regions [121]. The nanocellulose can act as a reinforcing agent for biopolymers and synthetic polymers [119], it being easily chemically compatible with the most diverse materials.

6.1 Nanocellulose as additive and coating of paper

Nanocellulose can be used to improve the properties of paper used in food packaging. For this application the types of nanocelluloses that are employed are MFC and NFC. MFC and NFC nanocelluloses have high capacity to form intra and interfibrillar bonds forming a dense network of fibrils. In addition to this characteristic, small voids are generates [121,122], MFC's have a high crystallinity index [123], which may have low permeability to air, oil and water.

The decrease in paper permeability is associated with increased bulk density [124]. Higher density occurs due to the higher contact surface of the nanocelluloses and the greater interaction and compression between them. The addition of nanocellulose in the production of paper with recycled fibers results in an apparent density increase of 7.5% to 25% and varies according to the type of paper to be recycled. In addition, as a consequence, porosity reduction occurs in relation to the paper without addition [125,126]. Similar result has been observed by testing different degrees of fibrillation and adding on paper, in which 15 passes increased apparent density by 5.6% and 35 and 50 passes increased apparent density by 9.43% [127]. Contrary to what be imagined, the increase in bulk density was not found proportional to the increase in refining in the production of nanocellulose by the mechanical method. The decrease in the paper density with the addition of high refining grade nanocelluloses is reported as due to the excess water in the intercellular spaces and the difficulty of the drainage with the press in the formation of the paper [128]. The increase in density and gain of mechanical and barrier properties are associated with a large amount of hydrogen bonds formed between nanocelluloses, which are very stable and require high energy to be broken [129].

Among the mechanical properties of the paper to which it has the greatest increments are the tensile strength and the bursting. There is a great variability in the value of the increment, which is related to the degree of processing of the nanocellulose as well as to the raw material to be used and the inherent binding capacity of the main fibers used [127,130]. Another factor that interferes is the amount of nanocellulose to be added, this being another very critical factor since no work so far has been able to measure exactly the degree of retention of the nanocellulose in the production of the wet mass and after the drying [131].

The increase of the surface loads, the decrease in size and the application of chemicals also help in the better adhesion between nanocelluloses, since the repellency between particles avoids the formation of agglomerates [131]. Naderi et al. [132] and Naderi et al. [133] working with mechanical processing with chemical additives, to produce phosphorylated NFC and carboxymethylated NFC, observed decrease in the size of nanocelluloses and increased surface charge resulting in increased mechanical properties of films. Further, they also observed phosphorylated NFC shown a reduction around 50% in oxygen passage.

There are reports of 2% increases in tensile properties [134] as well as increases above 100% [135,136] requiring caution in comparing the results. He et al. [127] found increases of 35%, 46%, 49%, 68% when adding 2% of nanocellulose processed with 15, 25, 35 and 50 passes.

In the case of recycled papers, significant increase in tensile strength and burst strength have been reported by the addition of NFC [125,126]. Viana et al. [126] have observed 97%, 133% and 104% increase in tensile, burst and tear resistance tests, by adding 10% of nanocellulose. Similarly, Delgado-Aguilar et al. [125] have reported improved workability of pulp in paper formation, reducing breakage due to improved fiber interlacing. Another advantage reported by these authors is preserving the structure of the fibers by using NFC instead of refining the fiber again. This would allow for a longer life and the possibility of reusing fibers more often. However, the main problem encountered in adding 4.5% NFC is reported in the drainability, which decreased by 36%, demonstrating that the removal of water from the paper would be more difficult.

Due to the reduction of porosity and increase in bulk density, the barrier properties are also altered [129]. The use of nanocelluloses as a surface coating on various base papers considerably reduces air permeability, but this effect is reduced to high relative humidity [129,135]. There are reports regarding air permeability determined by the Gurley method, which was reduced from 60% [137] to 280% [138]. Belbekhouche et al. [139] prepared films with CNC and NFC have concluded that the permeability of CNC films was much higher because of the nature of the pores formed in these films. On the other hand, Visanko et al. [140] have observed high oxygen resistance at 80% moisture in films produced with CNC, contrary to what was observed by Cozzolino et al. [141] who found a 20 times greater oxygen transmission in MFC films when compared to dry film.

According to Lagaron et al. [142] the effect of plasticization of cellulosic materials occurs, in high humidity causing a low cohesion in the film at high moisture content. Similar results are also reported by Lengowski et al. [143] and Wang et al. [144], which justify the loss in resistance of MFC films due to loss and absorption of water in the amorphous region of the nanocellulose.

Another way of improving the mechanical and barrier properties of papers is through the application of a thin superficial layer of nanocellulose [81,129], replacing the starch layer and sizing agents that make the coating to improve the hydrophobicity of the papers[131]. According to Kiviranta [145] all papers that are used for food packaging have a coating on their surface to improve their absorption properties. Fig.6 shows the main mechanisms that coatings aim to improve. The ability to absorb liquids and gases is directly related to porosity and paper density. For MFC based films the porosity can be modified by drying from different solvents, creating an adjustable feature that provides an advantage over the melt-formed plastics [81]. Henriksson et al. [146] were able to modify the porosity of water-dried MFC-based films from 28% up to 40% with films dried from solvents such as methanol and acetone. In addition, when used as a coat of MFC as a coating on paper, air permeability was found to reduce by 10% as a consequence of the reduction of surface porosity [121].

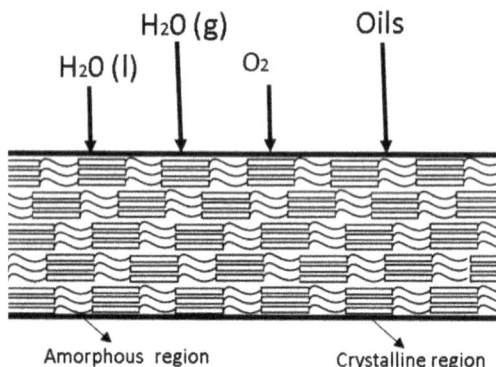

Figure 6. Main mechanisms that coatings are designed to improve.

6.2 Synthetic polymers

Because nanocellulose is a hydrophilic polymer while most polymer matrices are hydrophobic, nanocellulose makes dispersion a challenge in the development of new materials [147]. In addition, two other factors deserve special mention in the use of nanocellulose in polymer matrices. First one is the high availability of hydroxyl groups, which leads to the formation of agglomerates [148]. The second is their low thermal degradation point, which should be considered when it is used as reinforcement especially in polymers such as polypropylene, high and low density polyethylene, which have low degradation point [149].

Leao et al. [149] state that the high availability of free hydroxyl groups on the surface of cellulose nanofibrils causes increased agglomeration in the hydrophobic matrix of polypropylene, which suggest that the addition of natural isoprene can reduce the amount of OH and ultimately the agglomeration of nanocelluloses.

Other chemical modifications have already been tested in order to improve the dispersion of nanocellulose with the chemical modification with the use of surfactant polyoxyethylene [150–152], drying techniques of nanocellulose [153], processing as Solid State shear pulverization method [154] and the microfibrillation of the cellulose through extrusion while homogenizing with the polymer matrix [155,156].

Iwamoto et al. [150] found a better dispersion of the nanocellulose in polypropylene, however the improvement in mechanical properties was achieved only when mixing maleic anhydride to the blend. The significant improvement of interfacial adhesive interaction was observed, leading to a high Young's modulus, maintaining the great plastic deformation.

Lin and Dufresne [151] observed that polyoxyethylene and polyethylene glycol grafted nanocrystals improved thermal stability, allowing high temperature (200 °C) processing avoiding nanoparticle degradation and providing good dispersion and compatibility between modified nanocrystals and matrix.

Ljungberg et al. [152] using surfactant in the surface modification of the CNC in the matrix of atactic polypropylene observed greater transparency and better dispersion. They concluded that the improvement of the mechanical properties is not restricted only to the amount of reinforcement, but is associated with the chemical interaction that takes place between the matrix and the reinforcement.

Recently, studies have shown that spray-freeze drying produces porous microstructures that are easy to disperse in the polymer and increases the interfacial contact area of the CNC with matrix polymer [153,157]. Another study with polypropylene (PP) nanocomposites containing spray-dried CNC, freeze-dried CNC, and spray-freeze-dried CNC revealed possibility of their preparation, which showed the best results of dispersion, while composites produced with spray-freeze-dried CNC exhibited high Young's modulus [158].

Another study by Iyer et al. [154] showed 69% increase in the Young's modulus in low density polyethylene by the addition of 10% CNC by the solid state shear pulverization method. Further, the authors observed a 53% increase, being the largest increases found in the mechanical properties in polypropylene matrices, without modifying/making compatible the CNC.

The addition of 12% NFC to high density polyethylene resulted in a 44.7% increase in tensile strength; however the composite fragility increased over the control [159]. Modification in nanocelluloses before addition to high density polyethylene by aliphatic esters provides greater dispensability, enhanced Young's modulus, tensile strength and coefficient of thermal expansion (CTE) compared to those of nanocomposites produced with unmodified nanocelluloses [160].

Studies have shown that the production of low density polyethylene based films and polycaprolactone top layer containing surface modified crystalline nanocelluloses with maleic anhydride and copper oxide to be the most promising material for improvement in antibacterial and antifungal activity, as well as an oxygen transmission rate, which was 16% lower compared to pure low density polyethylene [161]. Although it is reported great difficulty in the compatibilization and homogenization of nanocellulose. Sapkota et al. [162] achieved good dispersion of the CNC without the use of organic solvents or compatibilized through a premix and subsequent melt, obtaining gains in the Young's modulus, maximum stress and decrease of the breaking stress in relation to the low density polyethylene without CNC .

In the case of poly (ethylene-co-vinyl acetate), addition of nanocellulose results in modified morphology of the cells during the process of foam expansion, since the nanocelluloses act as nucleating agents. The authors observed occurrence of the most efficient nucleation with 1% CNC along with uniformity of cell size. The mechanical properties of the compressive strength of the reinforced foams were found to be superior to those of the pure ethylene vinyl alcohol (EVA) foam due to the decrease in the size of the cells. Nanocomposites of ethylene vinyl alcohol copolymer (EVOH) containing CNC have been prepared by electrospinning and dissolution precipitation before matrix melting [163]. An improvement in the thermal properties of nanocomposites was observed in relation to pure EVOH [164].

Suzuki et al. [156] and Suzuki et al. [155] have reported processing of homogenized kraft pulp and polypropylene (PP) in a twin screw extruder in order to obtain nanocellulose and melt the polypropylene in a single step. The compatibilization of PP with cellulose was only possible with the use of polypropylene maleic anhydride or with cationic polymer having a primary amino group. Improvements in mechanical properties were found and thermal distortion temperature under a load of 1.82 MPa have been observed.

Polystyrene (PS) reinforced with CNC showed increase in the properties of storage modulus, crystallinity index, thermal decomposition temperature, tensile strength and in hydrophobicity [165,166]. The addition of 7% of CNC in PS by electrospinning increased the tensile strength by 170%, while the ductility decreased, while contact angle increased from 120 ° to 138 ° with the addition of 1% of CNC [165]. However the reinforcement with NFC produced by the TEMPO method produced a nanocomposite with high optical transparencies and its tensile strengths, elastic moduli and thermal dimensional stabilities increased with increasing NFC content [167].

Polyvinyl chloride (PVC) reinforced with CNC showed good dispersion and resulted in improved tensile strength and biodegradability of nanocomposites, but reduced thermal stability compared to pure PVC [168]. The chemical modification of CNC with silane resulted in its better interaction with the matrix along with reduction in voids and significant improvement in tensile strength of nanocomposites without chemical modification[169].

Polymethyl methacrylate (PMMA) reinforced with styrene-modified CNC showed an initial CNC degradation temperature was increased 50 °C after the modification. Also, the breaking strength and the elongation at break of the composites were improved and the transparency remained close to that of the pure PMMA [170]. of PMMA with CNF resulted in translucent nanocomposites, with increased thermal stability, glass transition temperature and the storage modulus. The Young's modulus and tensile strength increased with increasing nanofiber content [171,172].

Films composed of polyethylene terephthalate (PET) and nanocellulose exhibit a higher Young's modulus (about 10 GPa) and lower elongation (5.1%) and lower oxygen permeability of the films than that of PET films [173]. According to Leao et al. [149] the increase of filler nanocellulose restricts the mobility of the polymer chain and this is directly reflected in the values of % elongation and Young's modulus.

It is reported by Nair et al. [174] that oxygen barrier of nanocellulose films of 21±1 μm thickness is 17±1 ml m^{-2}day^{-1}, a value that is competitive with films of synthetic polymers of the same thickness, such as ethylene vinyl alcohol (EVOH) (3–5 ml·m^{-2}·day^{-1}) and polyester- oriented, coated polyvinylidene chloride (PVdC) films (9–15 ml·m^{-2}·day^{-1}).

Mascheroni et al. [175] working with PET films observed a high mechanical resistance when they applied a CNC coating on these films. This property was found to show higher increase when the CNC was produced through ammonium persulfate (APS) rather than sulfuric acid probably due to highly negative surface charge, resulting in strong carboxylic bonds between the CNCs.

Yousefian and Rodrigues [176] using 3% CNCs without modification in nylon 6 through the melting process, observed an increase of 41%, 23% and 11% in flexion strength, Young's modulus and tensile strength respectively. Rahimi and Otaigbe [177] observed an improvement in fluency and viscoelastic properties by adding 2% of CNC in nylon 6. The authors justified this result due to the formation of a percolation network of CNCs in the matrix. Lee et al. [178] observed 55% increase in tensile strength in polyamide-6 based nanocomposites containing CNF treated with amino silane.

Panaitescu et al. [179] used nylon 11 as matrix and 1% CNCs as reinforcement. They observed 25% increase in Young's modulus and 10% elongation. The authors report a good thermal stability (loss of mass less than 1% up to 270 °C) and justify this stability as due to increased crystallinity of the nanocomposite due to the addition of CNC.

Another study has reported reinforcement of polycarbonate (PC) with 18% NFC to prepare nanocomposites [180]. A 30% increase in the tensile modulus was observed while nanocomposites with 3% CNC reinforcement resulted in a good interaction between the PC carbonyl groups and the hydroxyl groups of the CNCs, Young's modulus to 27.3% increase, 30.6% increase in tensile strength and 3.3% in thermal stability [181]. However, another study has reported reduction of Tg and increase in the rate of depolymerization with increase of the CNC content in nanocomposites of polycarbonate prepared by master batch process [182].

6.3   Biopolymers based nanocomposites

The encouragement of the use of packaging that is not derived from petroleum has led to increased research in the area of biopolymers by institutes, universities and industries. However, these materials have low mechanical and barrier properties to water and oxygen [183]. In this context, reinforcement with nanomaterials becomes an interesting alternative.

CNC-reinforced Polylactic Acid (PLA) films produced by fusion extrusion method obtained higher tensile strength and temperature resistance, but decrease of elongation with increase of CNC content. However, it is still worth noting that the results demonstrated potential for food protection from oxidation and moisture reactions [184,185]. The surface modification with surfactant allows a better dispersion in the PLA matrix, besides the improvement of the barrier properties to oxygen and water. Both composites presented the migration level below the overall migration limits in both non-polar and polar simulants, suggesting the possibility to use these systems in food packaging applications [186].

The addition of 5% NFC to PLA also improved the viscoelastic properties of the films, especially at higher temperatures. However, the authors report difficulties in obtaining a homogeneous dispersion in the composite with this type of nanocellulose [187].

Another study with biopolymer PLA film of 0.25 μm thick coated with a 0.8 μm thick film of nanocelluloses has exhibited high oxygen barrier properties [188].The nanocellulose film reduced the oxygen permeability from 7,4 $ml \cdot m^{-2} \cdot day^{-1} \cdot kPa^{-1}$ for pure a PLA film to 9,9 $\times$ 10^{-4} $ml \cdot m^{-2} \cdot day^{-1} \cdot Pa^{-1}$ for PLA after the coating [188].

Nanocomposites prepared with CNC modified with silane and PLA had insignificant impact on PLA crystallinity [189]. However, the modification provided active nucleation sites and accelerated the PLA nucleation rate. Values of Young's modulus and tensile strength of the nanocomposite films prepared with 1% silane modified CNC were found to be 20% higher than those of the films prepared with unmodified CNC [189].

Dufresne and Vignon [190] and Dufresne et al. [191] prepared potato starch-based films and concluded that increase in the Young's Modulus is linear with the proportion of MFC added. Improvement of the mechanical properties of films with cassava starch was observed this seemed to depend on the degree of fibrillation of the MFC [135]. The improvement of the barrier properties with the addition of nanocellulose is attributed to the increase in the tortuosity of the way to go by the permeants [192]. One of the great difficulties of producing starch-based films is that it renders the film brittle. This fact is improved with the addition of plasticizers, however the barrier properties are impaired [193]. Films produced with starch and CNC reinforced showed good mechanical properties [194] with the reduction of the expansion and solubility in water [195].

Follain et al. [196] observed the improvement in water vapor resistance when adding CNC to poly (ε-caprolactone) films and attributed this improvement to the tortuosity of the path that the vapor needs to go through with the addition of nanocellulose. Saxena and Ragauskas [197] found a reduction in water vapor transmission of 74% when adding CNC in xylan film.

Dhar et al. [198] were able to improve the oxygen barrier properties on poly (3-hydroxybutyrate) films by adding CNC. Abdollani et al. [199] improved the wettability of alginate films with the addition of CNC besides an increase in contact angle from 41 ° to 71 ° by the addition of 5% CNC.

Thermoplastic starch-polyvinyl alcohol composite films reinforced with nanocellulose showed a 28% gain in traction property with the addition of 5% cellulose nanofibers, since water absorption was related to the content of glycerol present in the plasticization [200].

Khan et al. [201] prepared films based on methylcellulose, vegetable oil, glycerol and Tween-80 and observed that the mechanical properties and waterproofing of the films were slightly improved with low additions of nanocellulose.

An improvement in water vapor barrier properties and mechanical properties was observed in films produced with chitosan with addition in the NFC [202]. These nanocellulose films, being a biodegradable polymer and possess antimicrobial and antifungal properties, can be used to prepare films or edible coatings that do not demand high flexibility or high barrier capacity, and therefore can be used for improving the shelf life of food. Similar results for the mechanical and barrier properties have been reported in films of sodium caseinate [203] and in alginate films [204] using CNC. Alginate based nanocomposite films also showed a significant improvement in the transparency of the films with 10% of CNC incorporation; however, the mechanical properties of these nanocomposite films were found to be very much lower than those of nanocomposites containing less than 5% CNC.

7.   Final considerations

Historically, food packaging was created looking for four main characteristics: containment, related to the ease of transportation; convenience, related to the ease it brings to the consumed; protection and preservation related to avoiding leaks or damaging food as well as protecting from entry of microorganisms that may reduce shelf life; and communication, responsible for passing information on food quality and preparatory guidelines. However, the demand of the modern consuming food with natural quality food, guaranteed safety, minimum processing, extended shelf life, low cost, ready for consumption concept, associated with packaging with good mechanical properties, barrier and with the environmental appeal made with that the food industry was looking for new materials to satisfy these new needs.

In this context the raw material used for the production of the packaging plays an important and decisive role in the quality of the food and in the period of validity. Nanotechnology then emerges as an ally of these new demands, opening up various possibilities of materials, such as nanocellulose in the context of improved packaging.

Advanced Applications of Polysaccharides and their Composites   Materials Research Forum LLC
Materials Research Foundations **73** (2020) 136-183   https://doi.org/10.21741/9781644900772-5

The use of nanocellulose in the most diverse polymers that are used for the production of food packaging points to the creation of ecologically less aggressive packaging with better barrier properties, mechanical properties, thermal resistance, flexibility and stability.

Acknowledgments

At the outset, the authors express their sincere thanks to the Editor of the book, Prof. Inamuddin for inviting us to contribute this Chapter. The authors place on record and appreciate the kind permission given by M/s. Elsevier Inc Publishers, Springer Nature to reproduce some of the figures and one table from their publications free of charges. One of the authors (KGS) would like to thank the PPISR, Bangalore-India with whom he is associated with presently for their encouragement and interest in this collaboration.

References

[1]   W. Patrick Noonan, C. Noonan, Legal requirements for "functional food" claims, Toxicol. Lett. 150 (2004) 19–24. https://dx.doi.org/10.1016/j.toxlet.2003.05.002.

[2]   R.T. Mendes, R. Vilarta, G.L. Gutierres, Qualidade de vida e cultura alimentar, Ipê Editorial, Campinas, 2009.

[3]   N.-S. Kwak, D.J. Jukes, Functional foods. Part 1: the development of a regulatory concept, Food Control. 12 (2001) 99–107. https://dx.doi.org/10.1016/S0956-7135(00)00028-1.

[4]   F.P. Moraes, L. Colla, Functional foods and nutraceuticals: definition, legislation and health benefits, Rev. Eletrônica Farmácia. 3 (2006) 109–122.

[5]   U.L. Opara, A. Mditshwa, A review on the role of packaging in securing food system: Adding value to food products and reducing losses and waste, African J. Agric. Res. 8 (2013) 2621–2630. https://doi.org/10.5897/AJAR2013.6931.

[6]   T.E. Quested, A.D. Parry, S. Easteal, R. Swannell, Food and drink waste from households in the UK, Nutr. Bull. 36 (2011) 460–467. https://doi.wiley.com/10.1111/j.1467-3010.2011.01924.x.

[7]   F. Wikström, H. Williams, Potential environmental gains from reducing food losses through development of new packaging - a life-cycle model, Packag. Technol. Sci. 23 (2010) 403–411. https://doi.wiley.com/10.1002/pts.906.

[8]   T.E. Quested, A.D. Parry, S. Esteal, R. Swannell, Food and drink waste from households in the UK, Nutr. Bull. 36 (2011) 460–467. https://doi.org/10.1111/j.1467-3010.2011.01924.x.

[9]   N. Jorge, Food packagings, Cultura Acadêmica, São Paulo, 2013.

[10]   P.E.M. Bernardo, S.A. Navas, L. Tieco, F. Murata, M.R. da S. Alcântara, Bisphenol A: Review on its use in the food packaging, exposure and toxicity, Rev. Inst. Adolfo Lutz. 74 (2015) 1–11.

[11]   H. Wang, L. Wang, Developing a bio-based packaging film from soya by-products incorporated with valonea tannin, J. Clean. Prod. 143 (2017) 624–633. https://doi.org/10.1016/j.jclepro.2016.12.064.

[12]   K.L. Yam, P.T. Takhistov, J. Miltz, Intelligent Packaging: Concepts and Applications, J. Food Sci. 70 (2005) R1–R10. http://doi.wiley.com/10.1111/j.1365-2621.2005.tb09052.x.

[13]   G.L. Robertson, Food packaging : principles and practice, CRC Press, Boca Raton, 2013.

[14]   K. Marsh, B. Bugusu, Food Packaging - Roles, Materials, and Environmental Issues, J. Food Sci. 72 (2007) R39–R55. https://doi.wiley.com/10.1111/j.1750-3841.2007.00301.x.

[15]   L.M. Oliveira, P.A.P.L. V Oliveira, Principais agentes antimicrobianos utilizados em embalagens plásticas (Revisão), Braz. J. Food Technol. 7 (2004) 161–165.

[16]   F. Wikström, H. Williams, K. Verghese, S. Clune, The influence of packaging attributes on consumer behaviour in food-packaging life cycle assessment studies - a neglected topic, J. Clean. Prod. 73 (2014) 100–108. https://doi.org/10.1016/j.jclepro.2013.10.042.

[17]   H. Kour, N.A.T. Wani, A. Malik, R. Kaul, H. Chauhan, P. Gupta, A. Bhat, J. Singh, Advances in food packaging – a review, Stewart Postharvest Rev. 9 (2013) 1–7. https://dx.doi.org/10.2212/spr.2013.4.7.

[18]   J. Shin, S.E.M. Selke, Food Packaging, in: S. Clark, S. Jung, B. Lamsal (Eds.), Food Process., John Wiley &amp Sons, Ltd, Chichester, UK, 2014: pp. 249–273. https://doi.wiley.com/10.1002/9781118846315.ch11.

[19]   W. Soroka, Illustrated Glossary of Packaging Terminology, 2nd ed., Institute of Packaging Professionals, Naperville, 2008.

[20]   A.L. Missio, B.D. Mattos, D. de F. Ferreira, W.L.E. Magalhães, D.A. Bertuol, D.A. Gatto, A. Petutschnigg, G. Tondi, Nanocellulose-tannin films: From trees to sustainable active packaging, J. Clean. Prod. 184 (2018) 143–151. https://doi.org/10.1016/j.jclepro.2018.02.205.

[21]   M.L. Rooney, ed., Active Food Packaging, Springer US, Boston, MA, 1995. https://doi.org/10.1007/978-1-4615-2175-4.

[22]   L. Vermeiren, F. Devlieghere, J. Debevere, Effectiveness of some recent antimicrobial packaging concepts., Food Addit. Contam. 19 Suppl (2002) 163–171.

[23]   C.E. Realini, B. Marcos, Active and intelligent packaging systems for a modern society, Meat Sci. 98 (2014) 404–419. https://doi.org/10.1016/j.meatsci.2014.06.031.

[24]   S. Yildirim, B. Röcker, M.K. Pettersen, J. Nilsen-Nygaard, Z. Ayhan, R. Rutkaite, T. Radusin, P. Suminska, B. Marcos, V. Coma, Active Packaging Applications for Food, Compr. Rev. Food Sci. Food Saf. 17 (2018) 165–199. https://doi.org/10.1111/1541-4337.12322.

[25]   D. Dainelli, N. Gontard, D. Spyropoulos, E. Zondervan-van den Beuken, P. Tobback, Active and intelligent food packaging: legal aspects and safety concerns, Trends Food Sci. Technol. 19 (2008) S103–S112. https://doi.org/10.1016/j.tifs.2008.09.011.

[26]   E.F. Beitzen-Heineke, N. Balta-Ozkan, H. Reefke, The prospects of zero-packaging grocery stores to improve the social and environmental impacts of the food supply chain, J. Clean. Prod. 140 (2017) 1528–1541. https://doi.org/10.1016/j.jclepro.2016.09.227.

[27]   J.P. Kerry, M.N. O'Grady, S.A. Hogan, Past, current and potential utilisation of active and intelligent packaging systems for meat and muscle-based products: A review, Meat Sci. 74 (2006) 113–130. https://doi.org/10.1016/j.meatsci.2006.04.024.

[28]   L. Vermeiren, F. Devlieghere, M. van Beest, N. de Kruijf, J. Debevere, Developments in the active packaging of foods, Trends Food Sci. Technol. 10 (1999) 77–86. https://doi.org/10.1016/S0924-2244(99)00032-1.

[29]   P. Zhu, Z. Lin, J.M. Goddard, Performance of photo-curable metal-chelating active packaging coating in complex food matrices, Food Chem. 286 (2019) 154–159. https://doi.org/10.1016/j.foodchem.2019.01.195.

[30]   M. Wrona, K. Bentayeb, C. Nerín, A novel active packaging for extending the shelf-life of fresh mushrooms (Agaricus bisporus), Food Control. 54 (2015) 200–207. https://doi.org/10.1016/j.foodcont.2015.02.008.

[31]   Z.A.N. Hanani, F.C. Yee, M.A.R. Nor-Khaizura, Effect of pomegranate (Punica granatum L.) peel powder on the antioxidant and antimicrobial properties of fish gelatin films as active packaging, Food Hydrocoll. 89 (2019) 253–259. https://doi.org/10.1016/j.foodhyd.2018.10.007.

[32]   J.H. Han, Innovations in food packaging, Academic Press, Cambridge, 2005.

[33]   V. Coma, Bioactive packaging technologies for extended shelf life of meat-based products, Meat Sci. 78 (2008) 90–103. https://doi.org/10.1016/j.meatsci.2007.07.035.

[34]   V.T. Nguyen, M.J. Gidley, G.A. Dykes, Potential of a nisin-containing bacterial cellulose film to inhibit Listeria monocytogenes on processed meats, Food Microbiol. 25 (2008) 471–478. https://doi.org/10.1016/j.fm.2008.01.004.

[35]   M. Özdemir, J.D. Floros, Active food packaging technologies, Crit. Rev. Food Sci. Nutr. 44 (2004) 185–193. https://doi.org/10.1080/10408690490441578.

[36]   P. Appendini, J.H. Hotchkiss, Review of antimicrobial food packaging, Innov. Food Sci. Emerg. Technol. 3 (2002) 113–126. https://doi.org/10.1016/S1466-8564(02)00012-7.

[37]   C.G. Otoni, P.J.P. Espitia, R.J. Avena-Bustillos, T.H. McHugh, Trends in antimicrobial food packaging systems: Emitting sachets and absorbent pads, Food Res. Int. 83 (2016) 60–73. https://doi.org/10.1016/j.foodres.2016.02.018.

[38]   R. Ahvenainen, Novel food packaging techniques, CRC Press, Boca Raton, 2003.

[39]   S. Özilgen, M. Özilgen, Kinetic Model of Lipid Oxidation in Foods, J. Food Sci. 55 (1990) 498–498. https://doi.org/10.1111/j.1365-2621.1990.tb06795.x.

[40]   J.N. Coupland, D.J. McClements, Lipid oxidation in food emulsions, Trends Food Sci. Technol. 7 (1996) 83–91. https://doi.org/10.1016/0924-2244(96)81302-1.

[41]   T. Waraho, D.J. McClements, E.A. Decker, Mechanisms of lipid oxidation in food dispersions, Trends Food Sci. Technol. 22 (2011) 3–13. https://doi.org/10.1016/j.tifs.2010.11.003.

[42]   J. Yi, J. Ning, Z. Zhu, L. Cui, E.A. Decker, D.J. McClements, Impact of interfacial composition on co-oxidation of lipids and proteins in oil-in-water emulsions: Competitive displacement of casein by surfactants, Food Hydrocoll. 87 (2019) 20–28. https://doi.org/10.1016/j.foodhyd.2018.07.025.

[43]   S.S. Okubanjo, S.M. Loveday, A.M. Ye, P.J. Wilde, H. Singh, Droplet-stabilized oil-in-water emulsions protect unsaturated lipids from oxidation, J. Agric. Food Chem. (2019) acs.jafc.8b02871. https://doi.org/10.1021/acs.jafc.8b02871.

[44]   J.M. Lorenzo, P.E.S. Munekata, B. Gómez, F.J. Barba, L. Mora, C. Pérez-
Santaescolástica, F. Toldrá, Bioactive peptides as natural antioxidants in food products
– A review, Trends Food Sci. Technol. 79 (2018) 136–147.
https://doi.org/10.1016/j.tifs.2018.07.003.

[45]   J.M. Lorenzo, M. Pateiro, R. Domínguez, F.J. Barba, P. Putnik, D.B. Kovačević,
A. Shpigelman, D. Granato, D. Franco, Berries extracts as natural antioxidants in meat
products: A review, Food Res. Int. 106 (2018) 1095–1104.
https://doi.org/10.1016/j.foodres.2017.12.005.

[46]   J.M. Lorenzo, P.E.S. Munekata, A.S. Sant'Ana, R.B. Carvalho, F.J. Barba, F.
Toldrá, L. Mora, M.A. Trindade, Main characteristics of peanut skin and its role for
the preservation of meat products, Trends Food Sci. Technol. 77 (2018) 1–10.
https://doi.org/10.1016/j.tifs.2018.04.007.

[47]   T. Slots, L.H. Skibsted, J.H. Nielsen, The difference in transfer of all-rac-α-
tocopherol stereo-isomers to milk from cows and the effect on its oxidative stability,
Int. Dairy J. 17 (2007) 737–745. https://doi.org/10.1016/j.idairyj.2006.09.010.

[48]   R.P.P. Fernandes, M.A. Trindade, J.M. Lorenzo, P.E.S. Munekata, M.P. de Melo,
Effects of oregano extract on oxidative, microbiological and sensory stability of sheep
burgers packed in modified atmosphere, Food Control. 63 (2016) 65–75.
https://doi.org/10.1016/j.foodcont.2015.11.027.

[49]   R. Battisti, N. Fronza, Á. Vargas Júnior, S.M. da Silveira, M.S.P. Damas, M.G.N.
Quadri, Gelatin-coated paper with antimicrobial and antioxidant effect for beef
packaging, Food Packag. Shelf Life. 11 (2017) 115–124.
https://doi.org/10.1016/j.fpsl.2017.01.009.

[50]   Z.A. Maryam Adilah, Z.A. Nur Hanani, Active packaging of fish gelatin films
with Morinda citrifolia oil, Food Biosci. 16 (2016) 66–71.
https://doi.org/10.1016/j.fbio.2016.10.002.

[51]   J.F. Martucci, L.B. Gende, L.M. Neira, R.A. Ruseckaite, Oregano and lavender
essential oils as antioxidant and antimicrobial additives of biogenic gelatin films, Ind.
Crops Prod. 71 (2015) 205–213. https://doi.org/10.1016/j.indcrop.2015.03.079.

[52]   L. Barbosa-Pereira, I. Angulo, J.M. Lagarón, P. Paseiro-Losada, J.M. Cruz,
Development of new active packaging films containing bioactive nanocomposites,
Innov. Food Sci. Emerg. Technol. 26 (2014) 310–318.
https://doi.org/10.1016/j.ifset.2014.06.002.

[53]   S. Min, J.M. Krochta, Ascorbic Acid-containing whey protein film coatings for control of oxidation, J. Agric. Food Chem. 55 (2007) 2964–2969. https://pubs.acs.org/doi/abs/10.1021/jf062698r.

[54]   M. Huber, J. Ruiz, F. Chastellain, Off-flavour release from packaging materials and its prevention: a foods company's approach., Food Addit. Contam. 19 Suppl (2002) 221–8. https://doi.org/10.1080/02652030110072074.

[55]   A.L. Brody, E.R. Strupinsky, L.R. Kline, Active packaging for food applications, Technomic Pub. Co, 2001.

[56]   A. do N. Oliveira, M.C.P. Martins, C.G. de Oliveira, M.L. de souza Lopes, J.C.S. de Mattos, P.J.P. ESPITIA, J.M. de A. Teixeira, K. Pereira, N.R. de Melo, Evaluation of an active flavouring film and its application in cooked ham, Brazilian J. Food Technol. 13 (2011) 299–305. https://www.ital.sp.gov.br/bj/artigos/html/busca/PDF/v13n4436a.pdf.

[57]   P. Kotler, K.L. Keller, Marketing management, Pearson, 2016.

[58]   B. Kuswandi, Y. Wicaksono, Jayus, A. Abdullah, L.Y. Heng, M. Ahmad, Smart packaging: sensors for monitoring of food quality and safety, Sens. Instrum. Food Qual. Saf. 5 (2011) 137–146. https://doi.org/10.1007/s11694-011-9120-x.

[59]   P. Butler, Smart packaging–intelligent packaging for food, beverages, pharmaceuticals and household products, Mater. World. 9 (2001) 11–13.

[60]   M. Vanderroost, P. Ragaert, F. Devlieghere, B. De Meulenaer, Intelligent food packaging: The next generation, Trends Food Sci. Technol. 39 (2014) 47–62. https://doi.org/10.1016/j.tifs.2014.06.009.

[61]   V. Manthou, M. Vlachopoulou, Bar-code technology for inventory and marketing management systems: A model for its development and implementation, Int. J. Prod. Econ. 71 (2001) 157–164. https://doi.org/10.1016/S0925-5273(00)00115-8.

[62]   C. Chen, A.C. Kot, H. Yang, A two-stage quality measure for mobile phone captured 2D barcode images, Pattern Recognit. 46 (2013) 2588–2598. https://doi.org/10.1016/j.patcog.2013.01.031.

[63]   K. Domdouzis, B. Kumar, C. Anumba, Radio-frequency identification (RFID) applications: A brief introduction, Adv. Eng. Informatics. 21 (2007) 350–355. https://doi.org/10.1016/j.aei.2006.09.001.

[64]   K. Fujisaki, Evaluation of 13.56 MHz RFID System Considering Communication Distance Between Reader and Tag, in: L. Barolli, F. Xhafa, N. Javaid, T. Enokido

Advanced Applications of Polysaccharides and their Composites   Materials Research Forum LLC
Materials Research Foundations **73** (2020) 136-183   https://doi.org/10.21741/9781644900772-5

(Eds.), Innov. Mob. Internet Serv. Ubiquitous Comput., Springer International Publishing, Cham, 2019: pp. 190–200. https://doi.org/10.1007/978-3-319-93554-6_17.

[65]   C.M. Roberts, Radio frequency identification (RFID), Comput. Secur. 25 (2006) 18–26. https://doi.org/10.1016/j.cose.2005.12.003.

[66]   R. Want, Enabling ubiquitous sensing with RFID, Computer (Long. Beach. Calif). 37 (2004) 84–86. https://ieeexplore.ieee.org/document/1297315.

[67]   A.S. Martínez-Sala, E. Egea-López, F. García-Sánchez, J. García-Haro, Tracking of returnable packaging and transport units with active RFID in the grocery supply chain, Comput. Ind. 60 (2009) 161–171. https://doi.org/10.1016/j.compind.2008.12.003.

[68]   R. Jedermann, L. Ruiz-Garcia, W. Lang, Spatial temperature profiling by semi-passive RFID loggers for perishable food transportation, Comput. Electron. Agric. 65 (2009) 145–154. https://doi.org/10.1016/j.compag.2008.08.006.

[69]   E. Abad, S. Zampolli, S. Marco, A. Scorzoni, B. Mazzolai, A. Juarros, D. Gómez, I. Elmi, G.C. Cardinali, J.M. Gómez, F. Palacio, M. Cicioni, A. Mondini, T. Becker, I. Sayhan, Flexible tag microlab development: Gas sensors integration in RFID flexible tags for food logistic, Sensors Actuators B Chem. 127 (2007) 2–7. https://doi.org/10.1016/j.snb.2007.07.007.

[70]   A. Vergara, E. Llobet, J.L. Ramírez, P. Ivanov, L. Fonseca, S. Zampolli, A. Scorzoni, T. Becker, S. Marco, J. Wöllenstein, An RFID reader with onboard sensing capability for monitoring fruit quality, Sensors Actuators B Chem. 127 (2007) 143–149. https://doi.org/10.1016/j.snb.2007.07.107.

[71]   S.A. Hogan, J.P. Kerry, Smart Packaging of Meat and Poultry Products, in: J. Kerry, P. Butler (Eds.), Smart Packag. Technol. Fast Mov. Consum. Goods, John Wiley & Sons, Ltd, Chichester, UK, 2008: pp. 33–59. https://doi.org/10.1002/9780470753699.ch3.

[72]   K.B. Biji, C.N. Ravishankar, C.O. Mohan, T.K. Srinivasa Gopal, Smart packaging systems for food applications: a review., J. Food Sci. Technol. 52 (2015) 6125–35. https://doi.org/10.1007/s13197-015-1766-7.

[73]   A.C. Pinheiro, M.A. Cerqueira, B.W.S. Souza, J. Martins, J.A. Teixeira, A.A. Vicente, Use of edible coatings/films for food applications, Bol. Biotecnol. (2010) 18–28. https://hdl.handle.net/1822/16725.

[74]   F.M. Pelissari, D.C. Ferreira, L.B. Louzada, F. dos Santos, A.C. Corrêa, F.K.V. Moreira, L.H. Mattoso, Starch-Based Edible Films and Coatings: An Eco-friendly

Alternative for Food Packaging, in: M.T.P.S. Clerici, M. Schmiele (Eds.), Starches Food Appl., Academic Press, Cambridge, 2019: pp. 359–420. https://doi.org/10.1016/B978-0-12-809440-2.00010-1.

[75]   S.S. Nallan Chakravartula, C. Cevoli, F. Balestra, A. Fabbri, M. Dalla Rosa, Evaluation of drying of edible coating on bread using NIR spectroscopy, J. Food Eng. 240 (2019) 29–37. https://doi.org/10.1016/j.jfoodeng.2018.07.009.

[76]   T. Giancone, E. Torrieri, P. Di Pierro, L. Mariniello, M. Moresi, R. Porta, P. Masi, Role of constituents on the network formation of hydrocolloid edible films, J. Food Eng. 89 (2008) 195–203. https://doi.org/10.1016/j.jfoodeng.2008.04.017.

[77]   S. Galus, J. Kadzińska, Food applications of emulsion-based edible films and coatings, Trends Food Sci. Technol. 45 (2015) 273–283. https://doi.org/10.1016/j.tifs.2015.07.011.

[78]   H.J. Park, Development of advanced edible coatings for fruits, Trends Food Sci. Technol. 10 (1999) 254–260. https://doi.org/10.1016/S0924-2244(00)00003-0.

[79]   H.Y. Erbil, N. Muftugil, Lengthening the postharvest life of peaches by coating with hydrophobic emulsions, J. Food Process. Preserv. 10 (1986) 269–279. https://doi.org/10.1111/j.1745-4549.1986.tb00025.x.

[80]   P. Rantamäki, V. Loimaranta, E. Vasara, J. Latva-Koivisto, H. Korhonen, J. Tenovuo, P. Marnila, Edible films based on milk proteins release effectively active immunoglobulins, Food Qual. Saf. (2019). https://doi.org/10.1093/fqsafe/fyy027.

[81]   E.C. Lengowski, E.A. Bonfatti Júnior, M.M.N. Kumode, M.E. Carneiro, K.G. Satyanarayana, Nanocellulose in the Paper Making, in: Inamuddin, S. Thomas, R.K. Mishra, A.M. Asiri (Eds.), Sustain. Polym. Compos. Nanocomposites, Springer International Publishing, Cham, 2019: pp. 1027–1066. https://doi.org/10.1007/978-3-030-05399-4_36.

[82]   K. Marsh, B. Bugusu, Food Packaging - Roles, materials, and environmental issues, J. Food Sci. 72 (2007) R39–R55. http://doi.wiley.com/10.1111/j.1750-3841.2007.00301.x.

[83]   Y. Teck Kim, B. Min, K. Won Kim, General Characteristics of Packaging Materials for Food System, in: J.H. Han (Ed.), Innov. Food Packag., Academic Press, Cambridge, 2014: pp. 13–35. https://doi.org/10.1016/B978-0-12-394601-0.00002-3.

[84]   Z. Berk, Food packaging, in: Z. Berk (Ed.), Food Process Eng. Technol., Academic Press, Cambridge, 2018: pp. 625–641. https://doi.org/10.1016/B978-0-12-812018-7.00027-0.

[85]   B. Oraikul, M.E. Stiles, Modified atmosphere packaging of fruits and vegetables, Ellis Horwood, West Sussex, 1991.

[86]   ASTM, C162-05(2015): Standard Terminology of Glass and Glass Products, West Conshohocken, 2015. https://www.astm.org/Standards/C162.htm.

[87]   B. Luo, M. Ma, M.-A. Zhang, J. Shang, C.-P. Wong, Composite glass-silicon substrates embedded with microcomponents for mems system integration, IEEE Trans. Components, Packag. Manuf. Technol. 9 (2019) 201–208. https://ieeexplore.ieee.org/document/8586943/.

[88]   S. Sacharow, R.C. Griffin, Principles of food packaging, 2nd ed., Avi Publishers, New Delhi, 1980.

[89]   S. Keller, Paper drying in the manufacturing process, in: G. Banik, I. Brückle (Eds.), Pap. Water A Guid. Conserv., Butterworth Heinemann, Oxiford, 2013: pp. 173–211.

[90]   S.A. Gunaratne, Paper, Printing and the Printing Press, Gazette. 63 (2001) 459–479. https://doi.org/10.1177/0016549201063006001.

[91]   C. Robusti, E.F. Viana, F. Ferreira Júnior, I. Gomes, L. Tognetta, O. Dos Santos, P. Dragoni, Paper, SENAI-SP, São Paulo, 2014.

[92]   A.P.M. Landim, C.O. Bernardo, I.B.A. Martins, M.R. Francisco, M.B. Santos, N.R. de Melo, A.P.M. Landim, C.O. Bernardo, I.B.A. Martins, M.R. Francisco, M.B. Santos, N.R. de Melo, Sustainability concerning food packaging in Brazil, Polímeros. 26 (2016) 82–92. http://dx.doi.org/10.1590/0104-1428.1897.

[93]   V. Katiyar, S.S. Gaur, A. k. Pal, A. Kumar, Properties of Plastics for Packaging Applications, in: S. Alavi, S. Thomas, K.P. Sandeep, N. Kalarikkal, J. Varghese, S. Yaragalla (Eds.), Polym. Packag. Appl., 1st Editio, CRC Press, Boca Raton, 2014: pp. 03–38.

[94]   L. Lebreton, B. Slat, F. Ferrari, B. Sainte-Rose, J. Aitken, R. Marthouse, S. Hajbane, S. Cunsolo, A. Schwarz, A. Levivier, K. Noble, P. Debeljak, H. Maral, R. Schoeneich-Argent, R. Brambini, J. Reisser, Evidence that the great pacific garbage patch is rapidly accumulating plastic, Sci. Rep. 8 (2018) 4666. https://doi.org/10.1038/s41598-018-22939-w.

[95]   K. Petersen, P. Væggemose Nielsen, G. Bertelsen, M. Lawther, M.B. Olsen, N.H. Nilsson, G. Mortensen, Potential of biobased materials for food packaging, Trends Food Sci. Technol. 10 (1999) 52–68. https://doi.org/10.1016/S0924-2244(99)00019-9.

[96]   A. Sorrentino, G. Gorrasi, V. Vittoria, Potential perspectives of bio-nanocomposites for food packaging applications, Trends Food Sci. Technol. 18 (2007) 84–95. https://doi.org/10.1016/j.tifs.2006.09.004.

[97]   A.F. Turbak, F.W. Snyder, K.R. Sandberg, Microfibrillated cellulose, a new cellulose product: properties, uses, and commercial potential, J. Appl. Polym. Sci. Appl. Polym. Symp.; (United States). 37 (1983).

[98]   G. Mitchell, F. Gaspar, A. Mateus, V. Mahendra, D. Sousa, Advanced Materials from Forests, in: L.M.T. Martínez, O.V. Kharissova, B.I. Kharisov (Eds.), Handb. Ecomater., Springer, Cham, Chan, 2019: pp. 1–24. https://doi.org/10.1007/978-3-319-48281-1_189-1.

[99]   A.K. Bharimalla, S.P. Deshmukh, N. Vigneshwaran, P.G. Patil, V. Prasad, Nanocellulose-polymer composites for applications in food packaging: current status, future prospects and challenges, Polym. Plast. Technol. Eng. 56 (2017) 805–823. https://doi.org/10.1080/03602559.2016.1233281.

[100] C. Salas, M. Hubbe, O.J. Rojas, Nanocellulose Applications in Papermaking, in: Z. Fang, R.L. Smith, X. Tian (Eds.), Prod. Mater. from Sustain. Biomass, Springer, Singapore, 2019: pp. 61–96. https://doi.org/10.1007/978-981-13-3768-0_3.

[101] S. Fujisawa, Y. Okita, H. Fukuzumi, T. Saito, A. Isogai, Preparation and characterization of TEMPO-oxidized cellulose nanofibril films with free carboxyl groups, Carbohydr. Polym. 84 (2011) 579–583. https://doi.org/10.1016/j.carbpol.2010.12.029.

[102] H. Sehaqui, M. Allais, Q. Zhou, L.A. Berglund, Wood cellulose biocomposites with fibrous structures at micro- and nanoscale, Compos. Sci. Technol. 71 (2011) 382–387. https://doi.org/10.1016/j.compscitech.2010.12.007.

[103] D. Klemm, F. Kramer, S. Moritz, T. Lindström, M. Ankerfors, D. Gray, A. Dorris, Nanocelluloses: A new family of nature-based materials, Angew. Chemie Int. Ed. 50 (2011) 5438–5466. https://doi.wiley.com/10.1002/anie.201001273.

[104] A. Rodríguez-Rojas, A. Arango Ospina, P. Rodríguez-Vélez, R. Arana-Florez, What is the new about food packaging material: A bibliometric review during 1996–2016, Trends Food Sci. Technol. 85 (2019) 252–261. https://doi.org/10.1016/j.tifs.2019.01.016.

[105] N.M. Julkapli, S. Bagheri, Developments in nano-additives for paper industry, J. Wood Sci. 62 (2016) 117–130. https://doi.org/10.1007/s10086-015-1532-5.

[106] Y. Habibi, L.A. Lucia, O.J. Rojas, Cellulose Nanocrystals: Chemistry, Self-Assembly, and Applications, in: L.D. Madsen, E.B. Svedberg (Eds.), Mater. Res. Manuf., Springer, Cham, Cham, 2010: pp. 3479–3500. https://doi.org/10.1021/cr900339w.

[107] K. Nelson, T. Retsina, M. Iakovlev, A. van Heiningen, Y. Deng, J.A. Shatkin, A. Mulyadi, American Process: Production of Low Cost Nanocellulose for Renewable, Advanced Materials Applications, in: L.D. Madsen, E.B. Svedber (Eds.), Mater. Res. Manuf., Springer, Cham, Cham, 2016: pp. 267–302. https://doi.org/10.1007/978-3-319-23419-9_9.

[108] E.C. Lengowski, G.I.B. Muniz de, S. Nisgoski, W.L.E. Magalhães, Cellulose acquirement evaluation methods with different degrees of crystallinity, Sci. For. Sci. 41 (2013) 185–194.

[109] T. Nishino, I. Matsuda, K. Hirao, All-cellulose composite, Macromolecules. 37 (2004) 7863–7867. https://doi.org/10.1021/ma049300h.

[110] S. Janardhnan, M.M. Sain, Isolation of cellulose microfibrils – An enzymatic approach, BioResources. 1 (2006) 176–188.

[111] S. Kalia, S. Boufi, A. Celli, S. Kango, Nanofibrillated cellulose: surface modification and potential applications, Colloid Polym. Sci. 292 (2014) 5–31. https://doi.org/10.1007/s00396-013-3112-9.

[112] J. Rojas, M. Bedoya, Y. Ciro, Current Trends in the Production of Cellulose Nanoparticles and Nanocomposites for Biomedical Applications, in: M. Poletto (Ed.), Cellul. - Fundam. Asp. Curr. Trends, InTech, London, 2015: pp. 193–228. https://doi.org/10.5772/61334.

[113] M. Rabello, Aditivação de Polímeros, Artliber, São Paulo, 2000.

[114] G. Davis, J.H. Song, Biodegradable packaging based on raw materials from crops and their impact on waste management, Ind. Crops Prod. 23 (2006) 147–161. https://doi.org/10.1016/j.indcrop.2005.05.004.

[115] Q. Chaudhry, L. Castle, Food applications of nanotechnologies: An overview of opportunities and challenges for developing countries, Trends Food Sci. Technol. 22 (2011) 595–603. https://doi.org/10.1016/j.tifs.2011.01.001.

[116] C. Silvestre, D. Duraccio, S. Cimmino, Food packaging based on polymer nanomaterials, Prog. Polym. Sci. 36 (2011) 1766–1782. https://doi.org/10.1016/j.progpolymsci.2011.02.003.

[117] P. Podsiadlo, S.-Y. Choi, B. Shim, J. Lee, M. Cuddihy, N.A. Kotov, Molecularly Engineered Nanocomposites: Layer-by-Layer Assembly of Cellulose Nanocrystals, Biomacromolecules. 6 (2005) 2914–2918. https://doi.org/10.1021/bm050333u.

[118] A. Khan, T. Huq, R.A. Khan, B. Riedl, M. Lacroix, Nanocellulose-Based Composites and Bioactive Agents for Food Packaging, Crit. Rev. Food Sci. Nutr. 54 (2014) 163–174. https://doi.org/10.1080/10408398.2011.578765.

[119] A.K. Bharimalla, S.P. Deshmukh, N. Vigneshwaran, P.G. Patil, V. Prasad, Nanocellulose-polymer composites for applications in food packaging: current status, future prospects and challenges, Polym. Plast. Technol. Eng. 56 (2017) 805–823. https://doi.org/10.1080/03602559.2016.1233281.

[120] E.C. Lengowski, G.I.B. de Muñiz, A.S. de Andrade, L.C. Simon, S. Nisgoski, Morphological, physical and thermal characterization of microfibrillated cellulose, Rev. Árvore. 42 (2018). http://dx.doi.org/10.1590/1806-90882018000100013 .

[121] K. Syverud, P. Stenius, Strength and barrier properties of MFC films, Cellulose. 16 (2009) 75–85. https://doi.org/10.1007/s10570-008-9244-2.

[122] H. Fukuzumi, T. Saito, T. Iwata, Y. Kumamoto, A. Isogai, Transparent and High Gas Barrier Films of Cellulose Nanofibers Prepared by TEMPO-Mediated Oxidation, Biomacromolecules. 10 (2009) 162–165. https://doi.org/10.1021/bm801065u.

[123] C. Aulin, S. Ahola, P. Josefsson, T. Nishino, Y. Hirose, M. Österberg, L. Wågberg, Nanoscale cellulose films with different crystallinities and mesostructures—their surface properties and interaction with water, Langmuir. 25 (2009) 7675–7685. https://doi.org/10.1021/la900323n.

[124] K.L. Spence, R.A. Venditti, O.J. Rojas, Y. Habibi, J.J. Pawlak, The effect of chemical composition on microfibrillar cellulose films from wood pulps: water interactions and physical properties for packaging applications, Cellulose. 17 (2010) 835–848. https://doi.org/10.1007/s10570-010-9424-8.

[125] M. Delgado-Aguilar, I. González, M.A. Pèlach, E. De La Fuente, C. Negro, P. Mutjé, Improvement of deinked old newspaper/old magazine pulp suspensions by means of nanofibrillated cellulose addition, Cellulose. 22 (2015) 789–802. https://doi.org/10.1007/s10570-014-0473-2.

[126] L.C. Viana, D.C. Potulski, G.I.B. de Muniz, A.S. de Andrade, E.L. da Silva, L.C. Viana, D.C. Potulski, G.I.B. de Muniz, A.S. de Andrade, E.L. da Silva, Nanofibrillated cellulose as an additive for recycled paper, CERNE. 24 (2018) 140–148. https://dx.doi.org/10.1590/01047760201824022518 .

[127] M. He, G. Yang, B.-U. Cho, Y.K. Lee, J.M. Won, Effects of addition method and fibrillation degree of cellulose nanofibrils on furnish drainability and paper properties, Cellulose. 24 (2017) 5657–5669. https://doi.org/10.1007/s10570-017-1495-3.

[128] H. Sehaqui, L.A. Berglund, Q. Zhou, BIOREFINERY: Nanofibrillated cellulose for enhancement of strength in high-density paper structures, Nord. Pulp Pap. Res. J. 28 (2013) 182–189. https://doi.org/10.3183/npprj-2013-28-02-p182-189.

[129] C. Aulin, M. Gällstedt, T. Lindström, Oxygen and oil barrier properties of microfibrillated cellulose films and coatings, Cellulose. 17 (2010) 559–574. https://doi.org/10.1007/s10570-009-9393-y.

[130] C. Salas, M. Hubbe, O.J. Rojas, Nanocellulose Applications in Papermaking, in: Z. Fang, R.L. Smith, S. Tian (Eds.), Prod. Mater. from Sustain. Biomass Resour., Springer, Singapore, Singapore, 2019: pp. 61–96. https://doi.org/10.1007/978-981-13-3768-0_3.

[131] M.A. Hubbe, A. Ferrer, P. Tyagi, Y. Yin, C. Salas, L. Pal, O.J. Rojas, Nanocellulose in thin films, coatings, and plies for packaging applications: a review, BioResources. 12 (2017) 2143–2233.

[132] A. Naderi, J. Sundström, T. Lindström, J. Erlandsson, Enhancing the properties of carboxymethylated nanofibrillated cellulose by inclusion of water in the pretreatment process, Nord. Pulp Pap. Res. J. 31 (2016) 372–378. https://doi.org/10.3183/npprj-2016-31-03-p372-378.

[133] A. Naderi, T. Lindström, G. Flodberg, J. Sundström, K. Junel, A. Runebjörk, C.F. Weise, J. Erlandsson, Phosphorylated nanofibrillated cellulose: production and properties, Nord. Pulp Pap. Res. J. 31 (2016) 20–29. https://doi.org/10.3183/npprj-2016-31-01-p020-029.

[134] T. Taipale, M. Österberg, A. Nykänen, J. Ruokolainen, J. Laine, Effect of microfibrillated cellulose and fines on the drainage of kraft pulp suspension and paper strength, Cellulose. 17 (2010) 1005–1020. https://doi.org/10.1007/s10570-010-9431-9.

[135] E.C. Lengowski, Formation and characterization of films with nanocellulose, Federal University of Paraná, 2016. https://doi.org/10.13140/RG.2.2.13458.86722.

[136] J. Lee, K. Sim, K. Sim, H.J. Youn, Strengthening effect of surface treatment of cellulose nanofibrils on aged paper, J. Korea Tech. Assoc. Pulp Pap. Ind. 48 (2016) 123. https://doi.org/10.7584/JKTAPPI.2016.12.48.6.123.

[137] V. Ottesen, K. Syverud, Ø.W. Gregersen, Mixing of cellulose nanofibrils and individual furnish components: Effects on paper properties and structure, Nord. Pulp Pap. Res. J. 31 (2016) 441–447. https://doi.org/10.3183/npprj-2016-31-03-p441-447.

[138] I. González, S. Boufi, M.A. Pèlach, M. Alcalà, F. Vilaseca, P. Mutjé, Nanofibrillated cellulose as paper additive in eucalyptus pulps, Bio Resources. 7 (2012) 5167–5180.

[139] S. Belbekhouche, J. Bras, G. Siqueira, C. Chappey, L. Lebrun, B. Khelifi, S. Marais, A. Dufresne, Water sorption behavior and gas barrier properties of cellulose whiskers and microfibrils films, Carbohydr. Polym. 83 (2011) 1740–1748. https://doi.org/10.1016/j.carbpol.2010.10.036.

[140] M. Visanko, H. Liimatainen, J.A. Sirviö, K.S. Mikkonen, M. Tenkanen, R. Sliz, O. Hormi, J. Niinimäki, Butylamino-functionalized cellulose nanocrystal films: barrier properties and mechanical strength, RSC Adv. 5 (2015) 15140–15146. https://doi.org/10.1039/C4RA15445B.

[141] C.A. Cozzolino, G. Cerri, A. Brundu, S. Farris, Microfibrillated cellulose (MFC): pullulan bionanocomposite films, Cellulose. 21 (2014) 4323–4335. https://doi.org/10.1007/s10570-014-0433-x.

[142] J.M. Lagaron, R. Catalá, R. Gavara, Structural characteristics defining high barrier properties in polymeric materials, Mater. Sci. Technol. 20 (2004) 1–7. https://doi.org/10.1179/026708304225010442.

[143] E.C. Lengowski, W.L.E. Magalhães, S. Nisgoski, G.I.B. de Muniz, K.G. Satyanarayana, M. Lazzarotto, New and improved method of investigation using thermal tools for characterization of cellulose from eucalypts pulp, Thermochim. Acta. 638 (2016) 44–51. https://doi.org/10.1016/j.tca.2016.06.010.

[144] Z. Wang, A.G. McDonald, R.J.M. Westerhof, S.R.A. Kersten, C.M. Cuba-Torres, S. Ha, B. Pecha, M. Garcia-Perez, Effect of cellulose crystallinity on the formation of a liquid intermediate and on product distribution during pyrolysis, J. Anal. Appl. Pyrolysis. 100 (2013) 56–66. https://doi.org/10.1016/j.jaap.2012.11.017.

[145] A. Kiviranta, Paperboard grades, in: H. Paulapuro (Ed.), Pap. Board Grade Papermak. Sci. Technol., Fapet Oy, Helsinki, 2000: pp. 54–72.

[146] M. Henriksson, L.A. Berglund, P. Isaksson, T. Lindström, T. Nishino, Cellulose nanopaper structures of high toughness, Biomacromolecules. 9 (2008) 1579–1585. https://doi.org/10.1021/bm800038n.

[147] M. Bengtsson, P. Gatenholm, K. Oksman, The effect of crosslinking on the properties of polyethylene/wood flour composites, Compos. Sci. Technol. 65 (2005) 1468–1479. https://doi.org/10.1016/j.compscitech.2004.12.050.

[148] A.S. Singha, V.K. Thakur, Fabrication and characterization of h. sabdariffa fiber-reinforced green polymer composites, Polym. Plast. Technol. Eng. 48 (2009) 482–487. https://doi.org/10.1080/03602550902725498.

[149] A.L. Leao, B.M. Cherian, S. Narine, M. Sain, S. Souza, S. Thomas, Applications for Nanocellulose in Polyolefins-Based Composites, in: S. Mohanty, S.K. Kayak, B.S. Kaith, S. Kalia (Eds.), Polym. Nanocomposites Based Inorg. Org. Nanomater., John Wiley & Sons, Inc., Hoboken, NJ, USA, 2015: pp. 215–228. https://doi.org/10.1002/9781119179108.ch7.

[150] S. Iwamoto, S. Yamamoto, S.-H. Lee, T. Endo, Mechanical properties of polypropylene composites reinforced by surface-coated microfibrillated cellulose, Compos. Part A Appl. Sci. Manuf. 59 (2014) 26–29. https://doi.org/10.1016/j.compositesa.2013.12.011.

[151] N. Lin, A. Dufresne, Physical and/or chemical compatibilization of extruded cellulose nanocrystal reinforced polystyrene nanocomposites, Macromolecules. 46 (2013) 5570–5583. https://doi.org/10.1021/ma4010154.

[152] N. Ljungberg, C. Bonini, F. Bortolussi, C. Boisson, L. Heux, J.Y. Cavaillé, New nanocomposite materials reinforced with cellulose whiskers in atactic polypropylene: effect of surface and dispersion characteristics, Biomacromolecules. 6 (2005) 2732–2739. https://doi.org/10.1021/bm050222v.

[153] V. Khoshkava, H. Ghasemi, M.R. Kamal, Effect of cellulose nanocrystals (CNC) on isothermal crystallization kinetics of polypropylene, Thermochim. Acta. 608 (2015) 30–39. https://doi.org/10.1016/j.tca.2015.04.007.

[154] K.A. Iyer, G.T. Schueneman, J.M. Torkelson, Cellulose nanocrystal/polyolefin biocomposites prepared by solid-state shear pulverization: Superior dispersion leading to synergistic property enhancements, Polymer (Guildf). 56 (2015) 464–475. https://doi.org/10.1016/j.polymer.2014.11.017.

[155] K. Suzuki, H. Okumura, K. Kitagawa, S. Sato, A.N. Nakagaito, H. Yano, Development of continuous process enabling nanofibrillation of pulp and melt compounding, Cellulose. 20 (2013) 201–210. https://doi.org/10.1007/s10570-012-9843-9.

[156] K. Suzuki, A. Sato, H. Okumura, T. Hashimoto, A.N. Nakagaito, H. Yano, Novel high-strength, micro fibrillated cellulose-reinforced polypropylene composites using a cationic polymer as compatibilizer, Cellulose. 21 (2014) 507–518. https://doi.org/10.1007/s10570-013-0143-9.

[157] V. Khoshkava, M.R. Kamal, Effect of drying conditions on cellulose nanocrystal (CNC) agglomerate porosity and dispersibility in polymer nanocomposites, Powder Technol. 261 (2014) 288–298. https://doi.org/10.1016/j.powtec.2014.04.016.

[158] V. Khoshkava, M.R. Kamal, Effect of cellulose nanocrystals (cnc) particle morphology on dispersion and rheological and mechanical properties of polypropylene/CNC nanocomposites, ACS Appl. Mater. Interfaces. 6 (2014) 8146–8157. https://doi.org/10.1021/am500577e.

[159] W. Huanhuan, L. Qian, H. Zhiqian, L. Yuanmei, F. Shenyuan, Preparation and characterization of hdpe/nano-cellulose fiber composites, Plast. Sci. Technol. 21 (2014) 23–41. https://en.cnki.com.cn/Article_en/CJFDTotal-SLKJ201405015.htm.

[160] H. Yano, H. Omura, Y. Honma, H. Okumura, H. Sano, F. Nakatsubo, Designing cellulose nanofiber surface for high density polyethylene reinforcement, Cellulose. 25 (2018) 3351–3362. https://doi.org/10.1007/s10570-018-1787-2.

[161] N. Đorđević, A.D. Marinković, P. Živković, D. V Kovačević, S. Dimitrijević, V. Kokol, P.S. Uskoković, Improving the packaging performance of low-density polyethylene with PCL/nanocellulose/copper(II)oxide barrier layer, Sci. Sinter. 50 (2002) 149–161. https://doi.org/10.2298/SOS1802149D.

[162] J. Sapkota, J.C. Natterodt, A. Shirole, E.J. Foster, C. Weder, Fabrication and Properties of Polyethylene/Cellulose Nanocrystal Composites, Macromol. Mater. Eng. 302 (2017) 1600300. https://doi.org/10.1002/mame.201600300.

[163] M.V.G. Zimmermann, M.P. da Silva, A.J. Zattera, R.M. Campomanes Santana, Effect of nanocellulose fibers and acetylated nanocellulose fibers on properties of poly(ethylene-co-vinyl acetate) foams, J. Appl. Polym. Sci. 134 (2017). https://doi.org/10.1002/app.44760.

[164] M. Martínez-Sanz, A. Lopez-Rubio, J.M. Lagaron, Nanocomposites of ethylene vinyl alcohol copolymer with thermally resistant cellulose nanowhiskers by melt compounding (I): Morphology and thermal properties, J. Appl. Polym. Sci. 128 (2013) 2666–2678. https://doi.org/10.1002/app.38433.

[165] S. Huan, L. Bai, G. Liu, W. Cheng, G. Han, Electrospun nanofibrous composites of polystyrene and cellulose nanocrystals: manufacture and characterization, RSC Adv. 5 (2015) 50756–50766. http://doi.org/10.1039/C5RA06117B.

[166] O.J. Rojas, G.A. Montero, Y. Habibi, Electrospun nanocomposites from polystyrene loaded with cellulose nanowhiskers, J. Appl. Polym. Sci. 113 (2009) 927–935. https://doi.org/10.1002/app.30011.

[167] S. Fujisawa, T. Ikeuchi, M. Takeuchi, T. Saito, A. Isogai, Superior Reinforcement Effect of TEMPO-Oxidized Cellulose Nanofibrils in Polystyrene Matrix: Optical, Thermal, and Mechanical Studies, Biomacromolecules. 13 (2012) 2188–2194. https://doi.org/10.1021/bm300609c.

[168] G. Kadry, A.E.F. El-Hakim, Effect of nanocellulose on the biodegradation, morphology and mechanical properties of polyvinylchloride/ nanocellulose nanocomposites, Res. J. Pharm. Biol. Chem. Sci. 6 (2015) 659–666.

[169] R. Sheltami, H. Kargarzadeh, I. Abdullah, Effects of Silane Surface Treatment of Cellulose Nanocrystals on the Tensile Properties of Cellulose-Polyvinyl Chloride Nanocomposite, Sains Malaysiana. 44 (2015) 801–810. https://doi.org/10.17576/jsm-2015-4406-05.

[170] Y. Yin, X. Tian, X. Jiang, H. Wang, W. Gao, Modification of cellulose nanocrystal via SI-ATRP of styrene and the mechanism of its reinforcement of polymethylmethacrylate, Carbohydr. Polym. 142 (2016) 206–212. https://doi.org/10.1016/j.carbpol.2016.01.014.

[171] H. Dong, Y.R. Sliozberg, J.F. Snyder, J. Steele, T.L. Chantawansri, J.A. Orlicki, S.D. Walck, R.S. Reiner, A.W. Rudie, Highly Transparent and Toughened Poly(methyl methacrylate) Nanocomposite Films Containing Networks of Cellulose Nanofibrils, ACS Appl. Mater. Interfaces. 7 (2015) 25464–25472. https://doi.org/10.1021/acsami.5b08317.

[172] F. Fahma, N. Hori, T. Iwata, A. Takemura, The morphology and properties of poly(methyl methacrylate)-cellulose nanocomposites prepared by immersion precipitation method, J. Appl. Polym. Sci. 128 (2012) 1–6. https://doi.org/10.1002/app.38312.

[173] S. Fujisawa, Y. Okita, T. Saito, E. Togawa, A. Isogai, Formation of N-acylureas on the surface of TEMPO-oxidized cellulose nanofibril with carbodiimide in DMF, Cellulose. 18 (2011) 1191–1199. https://doi.org/10.1007/s10570-011-9578-z.

[174] S.S. Nair, J. Zhu, Y. Deng, A.J. Ragauskas, High performance green barriers based on nanocellulose, Sustain. Chem. Process. 2 (2014) 23. https://doi.org/10.1186/s40508-014-0023-0.

[175] E. Mascheroni, R. Rampazzo, M.A. Ortenzi, G. Piva, S. Bonetti, L. Piergiovanni, Comparison of cellulose nanocrystals obtained by sulfuric acid hydrolysis and ammonium persulfate, to be used as coating on flexible food-packaging materials, Cellulose. 23 (2016) 779–793. https://doi.org/10.1007/s10570-015-0853-2.

[176] H. Yousefian, D. Rodrigue, Effect of nanocrystalline cellulose on morphological, thermal, and mechanical properties of Nylon 6 composites, Polym. Compos. 37 (2016) 1473–1479. https://doi.org/10.1002/pc.23316.

[177] S. Kashani Rahimi, J.U. Otaigbe, Polyamide 6 nanocomposites incorporating cellulose nanocrystals prepared by In situ ring-opening polymerization: Viscoelasticity, creep behavior, and melt rheological properties, Polym. Eng. Sci. 56 (2016) 1045–1060. https://doi.org/10.1002/pen.24335.

[178] J.-A. Lee, M.-J. Yoon, E.-S. Lee, D.-Y. Lim, K.-Y. Kim, Preparation and characterization of cellulose nanofibers (CNFs) from microcrystalline cellulose (MCC) and CNF/polyamide 6 composites, Macromol. Res. 22 (2014) 738–745. https://doi.org/10.1007/s13233-014-2121-y.

[179] D.M. Panaitescu, A.N. Frone, C. Nicolae, Micro- and nano-mechanical characterization of polyamide 11 and its composites containing cellulose nanofibers, Eur. Polym. J. 49 (2013) 3857–3866. https://doi.org/10.1016/j.eurpolymj.2013.09.031.

[180] S. Panthapulakkal, M. Sain, Preparation and characterization of cellulose nanofibril films from wood fibre and their thermoplastic polycarbonate composites, Int. J. Polym. Sci. 2012 (2012) 1–6. http://dx.doi.org/10.1155/2012/381342.

[181] W. Xu, Z. Qin, H. Yu, Y. Liu, N. Liu, Z. Zhou, L. Chen, Cellulose nanocrystals as organic nanofillers for transparent polycarbonate films, J. Nanoparticle Res. 15 (2013) 1562. https://doi.org/10.1007/s11051-013-1562-0%0A.

[182] M. Mariano, N. El Kissi, A. Dufresne, Melt processing of cellulose nanocrystal reinforced polycarbonate from a masterbatch process, Eur. Polym. J. 69 (2015) 208–223. https://doi.org/10.1016/j.eurpolymj.2015.06.007.

[183] C. Gómez H., A. Serpa, J. Velásquez-Cock, P. Gañán, C. Castro, L. Vélez, R. Zuluaga, Vegetable nanocellulose in food science: A review, Food Hydrocoll. 57 (2016) 178–186. https://doi.org/10.1016/j.foodhyd.2016.01.023.

[184] K. Oksman, A.P. Mathew, D. Bondeson, I. Kvien, Manufacturing process of cellulose whiskers/polylactic acid nanocomposites, Compos. Sci. Technol. 66 (2006) 2776–2784. https://doi.org/10.1016/j.compscitech.2006.03.002.

[185] M.P. Arrieta, E. Fortunati, F. Dominici, E. Rayón, J. López, J.M. Kenny, Multifunctional PLA–PHB/cellulose nanocrystal films: Processing, structural and thermal properties, Carbohydr. Polym. 107 (2014) 16–24. https://doi.org/10.1016/j.carbpol.2014.02.044.

[186] E. Fortunati, M. Peltzer, I. Armentano, L. Torre, A. Jiménez, J.M. Kenny, Effects of modified cellulose nanocrystals on the barrier and migration properties of PLA nano-biocomposites, Carbohydr. Polym. 90 (2012) 948–956. https://doi.org/10.1016/j.carbpol.2012.06.025.

[187] M. Jonoobi, J. Harun, A.P. Mathew, K. Oksman, Mechanical properties of cellulose nanofiber (CNF) reinforced polylactic acid (PLA) prepared by twin screw extrusion, Compos. Sci. Technol. 70 (2010) 1742–1747. https://doi.org/10.1016/j.compscitech.2010.07.005.

[188] H. Fukuzumi, T. Saito, S. Iwamoto, Y. Kumamoto, T. Ohdaira, R. Suzuki, A. Isogai, Pore size determination of tempo-oxidized cellulose nanofibril films by positron annihilation lifetime spectroscopy, Biomacromolecules. 12 (2011) 4057–4062. https://doi.org/10.1021/bm201079n.

[189] A. Pei, Q. Zhou, L.A. Berglund, Functionalized cellulose nanocrystals as biobased nucleation agents in poly(l-lactide) (PLLA) – Crystallization and mechanical property effects, Compos. Sci. Technol. 70 (2010) 815–821. https://doi.org/10.1016/j.compscitech.2010.01.018.

[190] A. Dufresne, M.R. Vignon, Improvement of starch film performances using cellulose microfibrils, Macromolecules. 31 (1998) 2693–2696. https://doi.org/10.1021/ma971532b.

[191] A. Dufresne, D. Dupeyre, M.R. Vignon, Cellulose microfibrils from potato tuber cells: Processing and characterization of starch-cellulose microfibril composites, J. Appl. Polym. Sci. 76 (2000) 2080–2092. https://doi.org/10.1002/(SICI)1097-4628(20000628)76:14%3C2080::AID-APP12%3E3.0.CO;2-U.

[192] M.D. Sanchez-Garcia, E. Gimenez, J.M. Lagaron, Morphology and barrier properties of solvent cast composites of thermoplastic biopolymers and purified cellulose fibers, Carbohydr. Polym. 71 (2008) 235–244. https://doi.org/10.1016/j.carbpol.2007.05.041.

[193] H.M.C. Azeredo, Fundamentos de estabilidade de alimentos, 2nd ed., Embrapa, Brazília, 2012.

[194] M.N. Anglès, A. Dufresne, Plasticized starch/tunicin whiskers nanocomposites. 1. structural analysis, Macromolecules. 33 (2000) 8344–8353. https://doi.org/10.1021/ma0008701.

[195] J.B.A. da Silva, F. V. Pereira, J.I. Druzian, Cassava starch-based films plasticized with sucrose and inverted sugar and reinforced with cellulose nanocrystals, J. Food Sci. 77 (2012) N14–N19. https://doi.org/10.1111/j.1750-3841.2012.02710.x.

[196] N. Follain, S. Belbekhouche, J. Bras, G. Siqueira, S. Marais, A. Dufresne, Water transport properties of bio-nanocomposites reinforced by Luffa cylindrica cellulose nanocrystals, J. Memb. Sci. 427 (2013) 218–229. https://doi.org/10.1016/j.memsci.2012.09.048.

[197] A. Saxena, A.J. Ragauskas, Water transmission barrier properties of biodegradable films based on cellulosic whiskers and xylan, Carbohydr. Polym. 78 (2009) 357–360. https://doi.org/10.1016/j.carbpol.2009.03.039.

[198] P. Dhar, U. Bhardwaj, A. Kumar, V. Katiyar, Poly (3-hydroxybutyrate)/cellulose nanocrystal films for food packaging applications: Barrier and migration studies, Polym. Eng. Sci. 55 (2015) 2388–2395. https://doi.org/10.1002/pen.24127.

[199] M. Abdollahi, M. Alboofetileh, M. Rezaei, R. Behrooz, Comparing physico-mechanical and thermal properties of alginate nanocomposite films reinforced with organic and/or inorganic nanofillers, Food Hydrocoll. 32 (2013) 416–424. https://doi.org/10.1016/j.foodhyd.2013.02.006.

[200] F. Fahma, Sugiarto, T.C. Sunarti, S.M. Indriyani, N. Lisdayana, Thermoplastic cassava starch-PVA composite films with cellulose nanofibers from oil palm empty fruit bunches as reinforcement agent, Int. J. Polym. Sci. 2017 (2017) 1–5. https://doi.org/10.1155/2017/2745721.

[201] R.A. Khan, S. Salmieri, D. Dussault, J. Uribe-Calderon, M.R. Kamal, A. Safrany, M. Lacroix, Production and properties of nanocellulose-reinforced methylcellulose-based biodegradable films, J. Agric. Food Chem. 58 (2010) 7878–7885. https://doi.org/10.1021/jf1006853.

[202] H.M.C. Azeredo, L.H.C. Mattoso, R.J. Avena-Bustillos, G.C. Filho, M.L. Munford, D. Wood, T.H. McHugh, Nanocellulose reinforced chitosan composite films as affected by nanofiller loading and plasticizer content, J. Food Sci. 75 (2010) N1–N7. https://doi.wiley.com/10.1111/j.1750-3841.2009.01386.x.

[203] M. Pereda, G. Amica, I. Rácz, N.E. Marcovich, Structure and properties of nanocomposite films based on sodium caseinate and nanocellulose fibers, J. Food Eng. 103 (2011) 76–83. https://doi.org/10.1016/j.jfoodeng.2010.10.001.

[204] M. Abdollahi, M. Alboofetileh, R. Behrooz, M. Rezaei, R. Miraki, Reducing water sensitivity of alginate bio-nanocomposite film using cellulose nanoparticles, Int. J. Biol. Macromol. 54 (2013) 166–173. https:doi.org/10.1016/j.ijbiomac.2012.12.016.

Advanced Applications of Polysaccharides and their Composites   Materials Research Forum LLC
Materials Research Foundations **73** (2020) 184-197   https://doi.org/10.21741/9781644900772-6

Chapter 6

Nanocellulose in Paper Making

Vismaya N. Kumar[a], Sharrel Rebello[b], Embalil Mathachan Aneesh[b], Raveendran Sindhu[c*],
Parameswaran Binod[c], Reshmy R[d], Eapen Philip[d] and Ashok Pandey[e]

[a]Cashew Export Promotion Council of India (CEPCI) Laboratory and Research Institute,
Kollam – 691 001, India

[b]Communicable Disease Research Laboratory (CDRL), St. Joseph's College, Irinjalakuda, India

[c]Microbial Processes and Technology Division, CSIR-National Institute of Interdisciplinary
Science and Technology (CSIR-NIIST), Trivandrum – 695 019, India

[d] Post Graduate and Research Department of Chemistry, Bishop Moore College,

Mavelikara -690 110, Kerala, India

[e]CSIR- Indian Institute for Toxicology Research (CSIR-IITR), 31 MG Marg,

Lucknow-226 001, India

sindhurgcb@gmail.com; sindhufax@yahoo.co.in

Abstract

Nanocellulose is cellulose fibrils with one of its dimensions in nanometer range. It shares specific properties of both cellulosic and nanoscale materials. The two main families of nanocellulose particles include cellulose nanocrystals (CNCs) and cellulose nanofibers (CNFs). Both families have found use in paper making with CNCs limited to surface coatings and CNFs have a wide range of use in paper making. Nanocellulose has gained great interest in the paper and pulp industry because of its abundant availability, renewable and eco-friendly nature. Nanopaper is advantageous over traditional pulp paper due to its high strength, optical transparency, thermal stability, smoothness, etc. It has been widely used as wet and dry strength agent and also as a coating to improve barrier properties of the paper. The barrier properties may be destroyed due to the hydrophilic nature of nanopaper, but it can be improved by surface modifications. This review addresses an overview of the currently adopted method in the pulp and paper industry, the role of nanotechnology in the industry, the classification of nanocellulose, and its application in paper making.

Keywords

Nanocellulose, Paper Making, Cellulose Nanofibrils, Cellulose Nanocrystals

Contents

1.   Introduction

The concept of paper making originated two millennia ago when the need for documenting major discoveries for a long term and communication between people at distant regions, arose. The different stages of the current paper making process are pulping, bleaching of the pulp and dewatering and production of the paper itself. Wood and other forest-based products have been serving the raw material in paper making so far. However, there has always been a search for novel raw materials for the production of paper. Paper is highly flexible to any surface modifications and is compatible with printing inks and coating colors that make the paper really important [1]. Functional additives (fillers, coating agents, retention aids, and wet- and dry- strength agents) and control additives (retention aids, drainage aids, pitch control agents, deformers, bacteriocides, and slimicides) have been added in order to modify the properties of the paper. The introduction of nanotechnology has opened its application in papermaking since the 1990s and it was able to produce paper of increased quality, biocompatibility

Advanced Applications of Polysaccharides and their Composites   Materials Research Forum LLC
Materials Research Foundations **73** (2020) 184-197   https://doi.org/10.21741/9781644900772-6

and low cost [2]. Nanotechnology and nanomaterials have gained significant importance in most modern industries because of its abundant availability, renewable nature, good quality, and efficiency. Paper making is one among them. The search for an alternative raw material for paper making has come to an end with the application of nanotechnology in the industry.

Cellulose, an abundant natural polymer has gained importance in nanotechnology due to its wide range of properties like its biocompatible nature, lower density, potent strength and mechanical properties and cost-effective too [3]. Cellulose has been used in food and pharmaceutical, textile and paint industry, etc. Its use has been limited due to the hygroscopic property and lack of melting point but has gained acceptance in the form of nanocellulose. Nanocellulose is derived from cellulose by reducing the size of cellulose fibers which due to its green nature has been widely used for a diverse range of applications. The biodegradability of nanocellulose has made it an inevitable material in the production of commercially valued biocompatible products.

2.   Classification of nanocellulose

Nanocellulose has been classified in a number of ways but the paper making industries have classified them into two main classes. The two main families of nanocellulose particles include nano-objects and nanostructures. Nano-objects are further subdivided into cellulose microcrystal (CMC) and nanostructures into cellulose nanocrystals (CNCs) and cellulose nanofibers (CNFs). But nanostructures have found use in paper making with CNCs limited to surface coatings and CNFs have a wide range of use in paper making. Nanocellulose is also produced by some bacteria and is called bacterial nanocellulose (BNC). Depending on the sources and extraction methods adopted, the synthesized nanocellulose may vary in morphology, crystallinity, particle size and many other properties [4-6]. Table 1 shows the classification of nanocelluloses, source of production, dimensions and the mode of production. CNCs and CNFs have been widely used in paper making industries. So this review focuses mainly on the application of CNCs and CNFs in the process of papermaking. Fig.1. shows the images of different types of nanocelluloses.

2.1   Cellulose nanocrystals

Cellulose nanocrystals (CNCs) or nanowhiskers are derived from the crystalline regions of cellulose microfibers which are extracted by appropriate action of the mechanical, chemical and enzymatic treatment [7]. CNCs are highly crystalline in nature and less flexible when compared to CNFs. These are rod-shaped structures of nanometer dimensions derived from cellulose. Mechanical treatments such as pressure-induced homogenization, ultrasonic treatments, cryocrushing, microfluidization, etc. apply shear forces to the bulky cellulose molecule and thereby separates the cellulose microfibrils forming CNCs. Chemical treatment method mainly utilizes hydrolysis of cellulose pulp by acid. Concentrated acid has the inherent ability to degrade the amorphous or disordered regions of the cellulose chains thereby retaining the crystalline part after acid hydrolysis. Concentrated hydrochloric acid and sulfuric acid has been widely used for the formation of whisker-like CNCs. Commonly adopted method for the production of cellulose nanocrystals is acid treatment. Though it is more eco-friendly than acid hydrolysis, production by enzyme hydrolysis is practiced on a low scale when compared to other techniques as it is a time-consuming process, provides low yield and the process conditions applied come out to be harsh on the molecule. There are many other methods for the production of CNCs as described by Xie et al., [8].

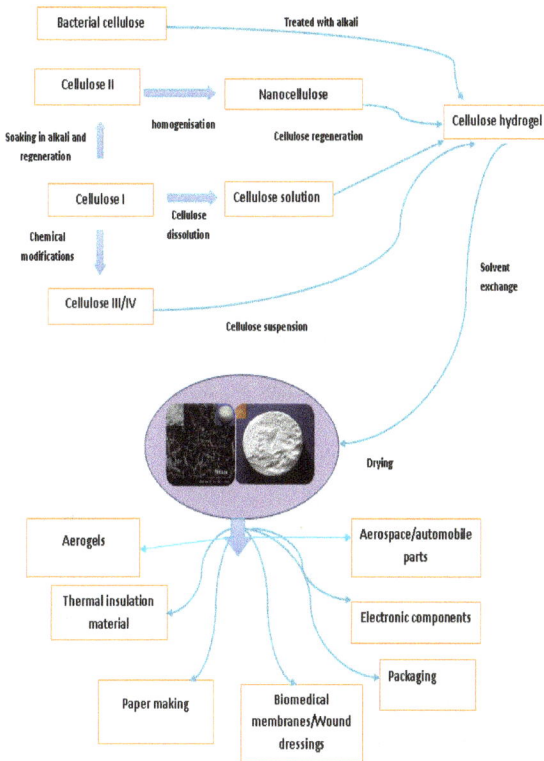

Fig. 1. Schematic representation of chemical modification and application of cellulose

Table 1*. Classification of nanocelluloses [42-44]*

Type of nanocellulose	Source of production	Diameter & length	Mode of production
Nanofibrillated cellulose(NFC)/ cellulose nanofibrils(CNFs)/microfibrillated cellulose(MFCs)/microfibrils	Wood, sugar beet, potato tuber, hemp, flax	Diameter - 5-60nm	Chemical or enzymatic pretreatment followed by intensive mechanical treatment
Nanocrystalline cellulose(NCC)/cellulose nanocrystals(CNCs)/whiskers	Wood, cotton, hemp, flax, wheat & rice straw, mulberry bark, ramie, avicel, tunicin, algae, and fungi	Diameter - 5-70nm Length - 100-250nm	Acid hydrolysis
Bacterial nanocellulose(BNC)/ bacterial cellulose/biocellulose/microbial cellulose	Bacteria like *Komagataeibacter, Zoogloea, Sarcina, Salmonella, Rhizobium, Pseudomonas, Escherichia, Agrobacterium, Aerobacter, Achromobacter, Azotobacter and Alcaligenes* using low molecular weight sugars and alcohol	Diameter - 20-100nm	From bacterial cultures

2.2   Cellulose nanofibers

Cellulose nanofibers (CNFs) also known as nano-fibrillated cellulose, cellulose microfibril, nano-fibrillar cellulose, etc. are long entangled nanosized materials derived from cellulose by various physical, chemical and enzymatic methods [4, 9]. Mechanical treatments include high-pressure grinding, homogenization, milling; chemical treatments include TEMPO oxidation and enzymatic treatments using biological enzymes like cellulase. Similar to the synthesis of cellulose nanocrystals, acid hydrolysis has also been adopted for the production of nano-fibrillar cellulose where the amorphous regions get dissolved leaving behind the fibers. Cellulolytic enzymes are also employed in its synthesis which cleaves the bulky molecule to simpler ones forming nanofibers. But enzymatic treatments are often combined with mechanical or chemical methods in order to improve yield and to make the process faster [10].

2.3   Bacterial nanocellulose

Bacterial nanocelluloses (BNCs) are a highly crystalline glucose polymer which was first reported to be produced by *Gluconacetobacter xylinus* previously named *Acetobacter xylinus* [11, 12]. Other Gram-negative bacteria like *Achromobacter, Pseudomonas, Agrobacterium, Azotobacter, Achromobacter, Rhizobium*, etc. As it is created in an eco-friendly manner by some microorganisms, it has gained much importance in the medical field, especially in tissue engineering as a scaffold, as a drug delivery route, as well as in other fields like biotechnological, pharmaceutical and food-based industries. Biomedical applications of bacterial cellulose are an upcoming area in the present medical world. Use of implants made out of BNCs has opened up a new era in the biomedical field [13]. It incorporates in its nanoscale structure some of the very beneficial properties like biologically compatible and functional nature, non-toxic and can be easily sterilized [14].

3.   Properties of nanocelluloses

A reduction in the size of materials to a nanoscale range imparts certain special properties to it, making it useful to a number of industrial applications. The inherent properties of cellulose macromolecule that undergo changes when resized are described below. The quality and efficiency of products of high market value can be enhanced by the application of nanotechnology as nanomaterials exhibit improved properties compared to large-sized ones.

3.1   Rheological properties

Rheological properties of nanocellulose include increased specific surface area as the size is reduced thereby the process of inter-particle interaction for various applications has become improved. But on drying, the specific surface area determined by gas adsorption isotherm may vary due to the irreversible aggregation of the nanoparticles [15]. It also exhibits high density due to the hydroxyl groups present on the surface. Due to the shear thinning behavior of cellulose nanoparticles, it becomes fluid when stirred at high shear rate.

3.2   Aspect ratio

Aspect ratio of nanoparticles is the ratio of its length to width and it depends on the source of cellulose and the production treatment conditions.

3.3   Thermal properties

Sulfate ester groups which form on the surface after acid hydrolysis by sulfuric acid imparts low thermal stability to CNCs compared to the other types.

3.4   Optical properties

As the dimensions of nanoparticles are very low compared to the wavelength of visible light, it cannot be optically viewed. Therefore, films made out of cellulose nanoparticles (CNCs/CNFs) incorporated with a transparent resin, are very transparent and can also be made semi-translucent by slow evaporation when in an aqueous medium. Fig. 2 shows the transparent nanopaper.

3.5   Barrier properties

Due to their high specific surface area, cellulose nanoparticles exhibit partially permeable nature as they form dense impermeable networks. It does not allow permeation of diffusing gases instead shows tortuosity or bending of the material. The barrier properties of cellulose nanoparticles make it useful as coatings for paper.

4.   Nanocellulose in papermaking

The easy availability, eco-friendly nature, renewability, and the enormous mechanical properties have made nanocellulose widely in use in the pulp and paper industry [16]. Conventionally, paper making process is done in three steps – pulping, bleaching and production of the product. Nanocelluloses have found use in paper making either as the substrate or as an additive or in coatings etc [2]. To avoid penetration of coating colors and ink, to make it's surface smooth, and to improve its optical properties and stability, fillers are added to the pulp. As nanotechnology has conquered the scenario, nano-based fillers (nanofillers) and pigments are used. Cellulose nanofibers are the most used among other nanocellulose materials in paper industry as nanopapers made out of CNFs serve all the mandatory qualities that a paper should possess like abundant wet and dry strength, optical transparency, high thermal stability, low thermal expansion, smoothness of the surface [1], tensile strength [17, 18] water barrier properties [19] in a study on agar mixed nanocellulose crystals. Fig. 3 shows nanopaper made from nanocelluloses.

CNFs were first used in papermaking by [20] as a coating material and found that it improves the strength of the final paper. In a study on the reduction in air permeability or gas resistance and also on the oil barrier properties and reported that due to the highly entangled fibers of nanocellulose, there is a reduction in the porosity of the paper which in-turn has lead to increased gas and oil barrier properties [21]. CNFs are also used in multilayer coatings where it acts as a binding agent that keeps the resin and the base paper together without penetrating into the paper [22]. The resulted paper exhibited high gas and water-vapor barrier properties and could be categorized under the high barrier category of papers used for food packaging and for modified atmosphere packaging (MAP). Rautkoski et al., used CNFs as a coating material for coating colors on paper but the problem faced by them was that such papers require more drying time and thus making the process a costly one [23].

Paper made using recycled fibers loses its strength and become brittle if subjected to increased refining process. Thus, in such cases, nanocellulose fibers find their use as a bonding agent [24]. Since CNFs are produced in the liquid state, it can be effectively and easily used as an additive at the beginning of paper making process or as a coating material at the end of the process [25]. Paper coatings are applied either to make the surface smooth and add additional properties and feature to the surface or to conceal the non-uniform nature of the surface of the paper. Coatings improve the smoothness, printability, surface energy, water retention and other barrier properties like gas, oil, etc. conventionally used coatings include mineral pigments like titanium dioxide, kaolin, zinc oxide, calcium carbonate, polymeric latex, and thickening agents [26]. But some of such compounds are harmful or toxic in nature and hence can be replaced with natural agents. With the emergence of nanotechnology has come to a solution that nanocelluloses can be used as coating agents in paper making [24].

The addition of CNFs to the pulping mixture improves the strength, makes it less porous, improves density and the increased specific surface area (due to the addition of CNFs) also increases the tensile strength and rigidity of the final paper [27]. Bardet and Bras, 2014 has made a review on the hybrid combinations of fillers/additives along with CNFs and has also listed out the various uses of CNFs in bulk paper applications [28]. The report shows that CNFs have improved wet and dry strength, tensile strength, water absorption, dimensional stability, act as a drainage aid, and has also optimized the production costs. The efficiency of the process depends on the rate at which CNFs have been retained in the base fibers. But it is difficult to monitor this as CNFs share a similar structural configuration to the base material used in papermaking. One better solution to this dilemma is to use fluorescent dyes or other tags to CNFs before its wet-end addition or processing. This can be used only for getting an initial knowledge on the retention rate of CNFs in the final paper and cannot be introduced when produced on a commercial level [29, 30]. In a study by Rautkoski et al., on the use of cellulose nanofibrils in coating colors, found that it can be better used as a thickening agent rather than as a binding agent [23]. Their study showed that the viscosity increased when latex was replaced with low dosages of CNFs. As CNFs could increase the viscosity to a great extent even at low concentration, it can be used as a thickening agent as they possess the ability to thicken up at very minute dosages and hence are not applicable as a binding agent.

One of the major drawbacks of the addition of CNFs in paper making slurry is that the drainability of the paper is affected. As Hubbe and Heitmann has mentioned, the drainability and square of the specific surface area are inversely related and reduced porosity of CNFs added papers can also lead to drainage resistance [31]. It is caused mainly due to the clogging of interfiber pores and reduction in its permeability as mentioned by Rantanen et al., [32]. A remedy to overcome this problem is the addition of retention aids to the slurry but the dosage of the retention aid used may adversely affect the mechanical properties of the final paper [25]. In a study by Ahola et al., CNFs were combined with Poly (amideamine) - epichlorohydrin (PAE) which improved the strength of the paper [33].

Another limitation faced in using nanocellulosic fibers in paper making is the high consumption of energy for its production and processing. There are many reports on the utilization of energy for the production processes and it shows that energy consumption is low when compared to the production of microfibrillated cellulose [34-37]. Researchers are working on reducing the amount of energy utilized in the various levels of the whole production system. One efficient method for the reduction of energy consumption had been described by Zheng [38] which shows that a combination of chemical or enzymatic treated and mechanical treatment may be a helpful idea to overcome this issue. Among the different chemical treatments tried, TEMPO-mediated oxidation was the most widely used by many researchers [35, 39, 40]. Other methods include periodate-chlorite oxidation, carboxymethylation, and acetylation apart from enzymatic treatment methods [41]. Poor retention of CNFs in the fibers is also a disadvantage reported against the use of nanocellulose in papermaking [28]. All the above-mentioned conditions of using nancellulose in papermaking have limited its usage to a small to pilot scale production rather than on a commercial scale.

Conclusions

In this chapter, we have discussed nanocelluloses, types of nanocelluloses, its properties, and its application in papermaking. Nanotechnology can be considered a great opportunity and a boon for the pulp and paper industry. However, the mode of production and the raw material from which the nanocellulose is produced will immensely affect its properties and application. Pulp and paper industry is benefited from the properties of nanocellulose like great tensile strength, biodegradability, barrier properties, optical transparency, coating properties, thermal stability, etc. All these properties could bring its level of application in paper and packaging to a height. But more research has to be done

Advanced Applications of Polysaccharides and their Composites
Materials Research Foundations **73** (2020) 184-197

Materials Research Forum LLC
https://doi.org/10.21741/9781644900772-6

at different angles before moving onto a commercial level of production. Albeit nanocelluloses have immense properties, there are also certain disadvantages faced when nanopapers are produced on a large scale like high energy consumption, low drainability, difficulties in obtaining a uniform surface. Researches have to be done more on reaching a solution to such problems in order for nanopapers to be commercialized.

Acknowledgment

Vismaya N Kumar acknowledges KSCSTE for funding (File No: KSCSTE/1056/2019-FSHP-LS). Raveendran Sindhu and Reshmy. R acknowledges DST for sanctioning projects under DST WOS-B scheme.

References

[1] P. Samyn, A. Barhoum, T. Öhlund, A. Dufresne, nanoparticles and nanostructured materials in papermaking, J. mater. Sci. 53(1) (2018) 146-184. https://doi.org/10.1007/s10853-017-1525-4

[2] E.C. Lengowski, E.A. Bonfatti Júnior, M.M.N. Kumode, M.E. Carneiro, K.G. Satyanarayana, Nanocellulose in the Paper Making. In: Inamuddin, Thomas S., Kumar Mishra R., Asiri A. (eds) Sustainable Polymer Composites and Nanocomposites. Springer, Cham. (2019), pp 1027-1066.

[3] K. P. Y. Shak, Y. L. Pang, S. K. Mah, Nanocellulose: Recent advances and its prospects in environmental remediation, Beilstein J. Nanotechnol. 9(1) (2018) 2479-2498. https://doi.org/10.3762/bjnano.9.232

[4] P. Phanthong, P. Reubroycharoen, X. Hao, G. Xu, A. Abudula, Nanocellulose : Extraction and application, Carbon Resourc. Conver. 1 (2018) 32-43. https://doi.org/10.1016/j.crcon.2018.05.004

[5] R. J. Moon, A. Martini, J. Nairn, J. Youngblood, A. Martini, J.Nairn, Chem Soc Rev Cellulose nanomaterials review : structure, properties, and nanocomposites, Chem. Soc. Rev. 40 (2011) 3941-3994. https://doi.org/10.1039/c0cs00108b

[6] Anwar, Z., Gulfraz, M., & Irshad, M. (2014). ScienceDirect Agro-industrial lignocellulosic biomass a key to unlock the future bio-energy : A brief review. *JRRAS*, 1–11. https://doi.org/10.1016/j.jrras.2014.02.003

[7] J. George, S. N. Sabapathi, Cellulose nanocrystals: synthesis, functional properties, and applications, Nanotech. Sci. application. 8 (2015) 45.

[8] H. Xie, H. Du, X. Yang, C. Si, Recent strategies in preparation of cellulose

nanocrystals and cellulose nanofibrils derived from raw cellulose materials, Int. J. Polym Sci. 2018 (2018) 25.

[9] M. Nasir, R. Hashim, O. Sulaiman, M. Asim, Nanocellulose: Preparation methods and applications. In Cellulose-Reinforced Nanofibre Composites, Woodhead Publishing. 2017 pp. 261-276.

[10] M. P. Menon, R. Selvakumar, S. Ramakrishna, Extraction and modification of cellulose nanofibers derived from biomass for environmental application, RSC Adv. 7(68) (2017) 42750-42773.

[11] F. Mohammadkazemi, M. Azin, A. Ashori, Production of bacterial cellulose using different carbon sources and culture media, Carbohydr. Polym. 117 (2015) 518–523. https://doi.org/10.1016/j.carbpol.2014.10.008

[12] S. Tanskul, K. Amornthatree, N. Jaturonlak, A new cellulose-producing bacterium, *Rhodococcus* sp. MI 2 : Screening and optimization of culture conditions, Carbohydr. Polym. 92(1) (2013) 421–428. https://doi.org/10.1016/j.carbpol.2012.09.017

[13] P. Gatenholm, D. Klemm, Bacterial nanocellulose as a renewable material for biomedical applications, MRS bull. 5(3) (2010) 208-213.

[14] A.F. Jozala, L.C. de Lencastre-Novaes, A.M. Lopes, V. de Carvalho Santos-Ebinuma, P.G. Mazzola, A. Pessoa-Jr, M.V. Chaud, Bacterial nanocellulose production and application: a 10-year overview, Appl. Microbiol. Biotechnol. 100(5) (2016) 2063-2072. https://doi.org/10.1007/s00253-015-7243-4

[15] A. Dufresne, Nanocellulose Processing Properties and Potential Applications, Curr. Forest. Rep. (2019) 1-14.

[16] V.S. Chauhan, S.K. Chakrabarti, Use of nanotechnology for high performance cellulosic and papermaking products, Cellulose chem. Tech. 46(5) (2012) 389.

[17] M. Ankerfors, Microfibrillated cellulose: Energy-efficient preparation techniques and key properties (Doctoral dissertation, KTH Royal Institute of Technology) (2012). https://www.diva-portal.org/smash/get/diva2:557668/FULLTEXT01.pdf

[18] J. A. Sirviö, A. Kolehmainen, H. Liimatainen, J. Niinimäki, O.E. Hormi, Biocomposite cellulose-alginate films: Promising packaging materials, Food chem. 151 (2014) 343-351. https://doi.org/10.1016/j.foodchem.2013.11.037

[19] J. P. Reddy, J. Rhim, Characterization of bionanocomposite films prepared with agar and paper-mulberry pulp nanocellulose, Carbohydr. Polym. 110 (2014) 480–488. https://doi.org/10.1016/j.carbpol.2014.04.056

[20] K. Syverud, P. Stenius, Strength and barrier properties of MFC films, Cellulose. 16(1), (2009) 75. https://doi.org/10.1007/s10570-008-9244-2

[21] C. Aulin, M. Gällstedt, T. Lindström, Oxygen and oil barrier properties of microfibrillated cellulose films and coatings, Cellulose. 17(3) (2010) 559-574. https://doi.org/10.1007/s10570-009-9393-y

[22] E. L. Hult, M. Iotti, M. Lenes, Efficient approach to high barrier packaging using microfibrillar cellulose and shellac, Cellulose. 17(3), (2010). 575-586. https://doi.org/10.1007/s10570-010-9408-8

[23] H. Rautkoski, H. Pajari, H. Koskela, A. Sneck, P. Moilanen, Use of cellulose nanofibrils (CNF) in coating colors, Nordic Pulp Paper Res. J. 30(3) (2015) 511-518.

[24] C. Salas, T. Nypelö, C. Rodriguez-Abreu, C. Carrillo, O.J. Rojas, Nanocellulose properties and applications in colloids and interfaces, Curr. Opin. Colloid Interface Sci. 19(5) (2014) 383-396. https://doi.org/10.1016/j.cocis.2014.10.003

[25] S. Boufi, I. González, M. Delgado-aguilar, Q. Tarrès, M.À. Pèlach, Nanofibrillated cellulose as an additive in papermaking process : A review, Carbohydr. Polym. 154 (2016) 151–166. https://doi.org/10.1016/j.carbpol.2016.07.117

[26] J. Brander, I. Thorn, (Eds.), Surface application of paper chemicals, Springer Science & Business Media, (2012).

[27] F.W. Brodin, Ø. Eriksen, Preparation of individualised lignocellulose microfibrils based on thermomechanical pulp and their effect on paper properties, Nordic Pulp Paper Res. J. 30(3) (2015) 443-451.

[28] R. Bardet, J. Bras, Cellulose nanofibers and their use in paper industry, In Handbook of Green Materials. (2014) 207-232.https://doi.org/10.1142/9789814566469

[29] R. Hollertz, V.L. Durán, P.A. Larsson, L. Wågberg, Chemically modified cellulose micro-and nanofibrils as paper-strength additives, Cellulose. 24(9) (2017) 3883-3899. https://doi.org/10.1007/s10570-017-1387-6

[30] C.R. Daniels, C. Reznik, C.F. Landes, Dye diffusion at surfaces: Charge matters, Langmuir. 26(7) (2010) 4807-4812. https://doi.org/10.1021/la904749z

[31] M.A. Hubbe, J.A. Heitmann, Review of factors affecting the release of water from cellulosic fibers during paper manufacture, Bio Resource. 2(3) (2007) 500-533.

[32] J. Rantanen, K. Dimic-Misic, J. Kuusisto, T.C. Maloney, The effect of micro and

nanofibrillated cellulose water uptake on high filler content composite paper properties and furnish dewatering, Cellulose. 22(6) (2015) 4003-4015.

[33] S. Ahola, M. Österberg, J. Laine, Cellulose nanofibrils—adsorption with poly (amideamine) epichlorohydrin studied by QCM-D and application as a paper strength additive, Cellulose. 15(2) (2008) 303-314.

[34] A. Isogai, T. Saito, H. Fukuzumi, TEMPO-oxidized cellulose nanofibers, nanoscale. 3(1) (2011) 71-85. https://doi.org/10.1039/c0nr00583e

[35] K. Syverud, G. Chinga-carrasco, J. Toledo, P.G. Toledo, A comparative study of *Eucalyptus* and *Pinus radiata* pulp fibers as raw materials for production of cellulose nanofibrils, Carbohydr. Polym. 84(3) (2011) 1033–1038. https://doi.org/10.1016/j.carbpol.2010.12.066

[36] Ø. Eriksen, K. Syverud, Ø. Eriksen, The use of microfibrillated cellulose produced from kraft pulp as strength enhancer in TMP paper, Nordic Pulp Paper Res. J. 23(3) (2008) 299–304.

[37] S. Josset, P. Orsolini, G. Siqueira, A. Tejado, P. Tingaut, T. Zimmermann, Energy consumption of the nanofibrillation of bleached pulp, wheat straw and recycled newspaper through a grinding process, Nordic Pulp Paper Res. J. 29(1) (2014) 167-175.

[38] H. Zheng, Production of fibrillated cellulose materials-Effects of pretreatments and refining strategy on pulp properties PhD Dessertation. (2014) 58. https://aaltodoc.aalto.fi/handle/123456789/13018

[39] Y. Qing, R. Sabo, J.Y. Zhu, U. Agarwal, Z. Cai, Y. Wu, A comparative study of cellulose nanofibrils disintegrated via multiple processing approaches, Carbohydr. Polym. 97(1) (2013). 226–234. https://doi.org/10.1016/j.carbpol.2013.04.086

[40] K. Kekäläinen, H. Liimatainen, M. Illikainen, T.C. Maloney, J. Niinimäki, The role of hornification in the disintegration behaviour of TEMPO-oxidized bleached hardwood fibres in a high-shear homogenizer, Cellulose. 21(3) (2014) 1163-1174. https://doi.org/10.1007/s10570-014-0210-x

[41] S.H. Osong, S. Norgren, P. Engstrand, Processing of wood-based microfibrillated cellulose and nanofibrillated cellulose, and applications relating to papermaking : A review, Cellulose. 23 (2016) 93-123. https://doi.org/10.1007/s10570-015-0798-5

[42] A. Blanco, M.C. Monte, C. Campano, A. Balea, N. Merayo, C. Negro, Nanocellulose for Industrial Use : Cellulose Nanofibers (CNF), Cellulose Nanocrystals (CNC), and Bacterial Cellulose (BC). Handbook of Nanomaterials

for Industrial Applications, Elsevier Inc. (2018). https://doi.org/10.1016/B978-0-12-813351-4.00005-5

[43] L. Brinchi, F. Cotana, E. Fortunati, J.M. Kenny, Production of nanocrystalline cellulose from lignocellulosic biomass : Technology and applications, Carbohydr. Polym. 94(1) (2013) 154–169. https://doi.org/10.1016/j.carbpol.2013.01.033

[44] D. Klemm, F. Kramer, S. Moritz, T. Lindström, M. Ankerfors, D. Gray, A. Dorris, Nanocelluloses: a new family of nature-based materials, Angew. Chem I. Ed. 50(24) (2011) 5438-5466.

Advanced Applications of Polysaccharides and their Composites   Materials Research Forum LLC
Materials Research Foundations 73 (2020) 198-241   https://doi.org/10.21741/9781644900772-7

Chapter 7

Starch-Based Composites and Their Applications

E.G. Okonkwo[a] *V.S. Aigbodion[a,b], E.T. Akinlabi[b,] C.C. Daniel-Mkpume[a]

[a]Department of Metallurgical and Materials Engineering, University of Nigeria, Nsukka, Nigeria

[b]Department of Mechanical Engineering Science, University of Johannesburg, P.O.BOX 524 Auckland park, South Africa

*victor.aigbodion@unn.edu.ng

Abstract

One of the major breakthroughs in biodegradable polymer composites is in the area of starch reinforced composites; materials with still-to-be explored potentials. Starch is a low cost biodegradable polymer with distinctive properties and shortcomings. Blending with other polymers have led to the development of an enhanced composite material with more advanced properties and varied areas of application. This paper aims at looking at the processing, applications and trend in the development of starch and starch blended composites. The effect of various additives and different class of polymers will be discussed while highlighting future trend in its developmental course.

Keywords

Starch, Composites, Starch-blend, Green Composites, Mechanical Properties, Microstructure, Application

Contents

1.   Introduction

Over the years, environmental impact has become a primary concern in manufacturing of engineering materials. Although this can be linked to pollution, degradation of environment and diminishing sources of fossil fuel. Sustainability of raw materials and variety in choice of materials is another way to look at it. It is no gain saying that polymer composites have come to stay. Infused with nanoparticles, they have continued to break new grounds in biomedicine, food packaging, tissue engineering among others. According to Google search trends, the term "green composite", is receiving a lot of research attention. This also seems to coincide with the number of publications on polymer composites as well. Continued research is still ongoing especially in the bid to enhance the properties of already existing composites [1–3]. Most of which are likened towards development of fully bio-compostable polymer composites; green composites. Green composites are composites in which the matrix and reinforcement are made up of biodegradable materials. According to [4], they are specific class of bio-composites formed by reinforcing a bio-based polymer with a biodegradable material such as natural fiber. Due to much improved fabrication techniques, better methods of improving the properties of natural fiber-reinforced polymer composites and infusion of nanomaterials in polymers, the developmental trend of green composites looks sustainable. Research on development and application of green composites is one that exceeds that of synthetic composite. History has shown that more than 3000 years ago, straws were mixed with clay to make bricks in Egypt. This practice of reinforcing clays with natural fibers and sticks for building mud houses is still common in developing countries in Africa, Asia etc. Although the trends have moved from using ceramic matrices like clay to polymers which not only extends the application of the ensuing material, the reinforcing phase has also received a facelift with the introduction of nanoparticles.

Over the last decades, biodegradable polymers especially for use as matrix materials in composites have continued to gain attention due to the ability to form non-toxic and eco-friendly materials. One of the most commonly available and low cost biodegradable polymer is starch. Being the main form of food storage for plants, its continual availability is unquestionable compared to polymers from fossil fuel. However, just like most biodegradable polymers, starch is brittle, has low glass transition temperature and limited mechanical strength. Nevertheless, its properties can be improved by blending with other polymers. Similarly, plasticization with water and or glycerol is another way of improving the properties of starch. With the prospect shown by nanomaterials and natural fibers, incorporation of nanofillers and natural fibers can improve mechanical properties, thermal stability, barrier properties especially against gases such as oxygen, carbon dioxide, moisture and flavours. Nanoclays such as montmorillonite (MMT) and kaolin due to their layered structure have shown potential in this aspect for not only starch but most other biodegradable and even synthetic polymers.

Introduction of nanomaterials in starch to form nanocomposites can extend the current areas of application of starch composites. Nanoparticles such as starch nanocrystal and polysaccharides nanomaterials forms hydrogen bonds with starch thus improving thermomechanical, thermal and physicochemical properties as well as interfacial bonding [5,6]. Starch is cheap, abundant and eco-friendly. These properties have made it possible for it to be applied in various areas. In food packaging, it is seen as a material for the future due to non-toxicity and improved barrier properties; in tissue engineering due to pores for cell regeneration and drug delivery due to control release rate. Starch composites like thermoplastic starch foams can be used as damping materials due to their ability to absorb shocks [6]. The field of application of starch and its composites is still growing. Increased understanding of the behavior of starch and possible methods of enhancing its properties brought about by increased research interest is a frontrunner. This article aims at giving a brief overview of starch and its composites, its properties, allied areas of applications while laying more emphasis on research efforts geared towards improving the properties of starch and its derivatives as well.

Advanced Applications of Polysaccharides and their Composites   Materials Research Forum LLC
Materials Research Foundations **73** (2020) 198-241   https://doi.org/10.21741/9781644900772-7

1.1   Starch as a green polymer

Starch is the primary form of carbohydrate found in roots, seeds, stalks of stable food crops like cassava, yam, rice, etc. It is the major food reserve of plants and provides up to 70% of energy found in diet of an average human being. Structurally, it is designed for ease of access to enzymes for energy forming degradation [7]. Besides from being a low cost polysaccharide and biodegradable, starch as a polymer is also known for its film forming ability which can be exploited for many applications unlike most petroleum based polymers. Conversely, it is hydrophilic, brittle and exhibits poor mechanical strength. Similar to starch is cellulose, a natural structural polymer designed for regularity of packing, chain stiffening and strong cohesion via hydrogen bonding. Cellulose is a structural polymer forming cell walls in plants. In the presence of water, celluose is difficult to melt at moderate temperatures. Just like starch, cellulose is gaining attention in various applications. According to [8–10], the application of cellulose in biomedicine continues to grow. Comparatively, the distinctive physicochemical and functional properties of starch makes it a material for not only today's but future technological applications. Economically, starch is cheaper than other natural polymers.

Structurally, starch is a semi-crystalline polymer consisting of two glucans (homopolymers) namely; amylose, amylopectin and minor components such as lipids and proteins [11]. Amylose is a linear polymer with α-1,4glycosidic linkages whereas amylopectin is a highly branched D-glucopyranose polymer with both α-1,6 and α-1,4 linkages. The ratio of amylose and amylopectin varies from plant to plant and depends on factors such as geographical location among others. The structure of starch granules depends on how these two polymers are distributed. For example, starch granules from waxy maize starch contains about 100% amylopectin whereas cassava starch contains 28 wt.-% amylose and 72 wt.-% amylopectin [5,11]. The ratio of amylose and amylopectin bears on the DS (degree of substitution) of starch derivatives. Starches with low amylose content exhibit higher DS on acetylation [12]. Higher amylose content in starch improves the film forming ability [13]. Fig. 1 below shows the structure of starch granule.

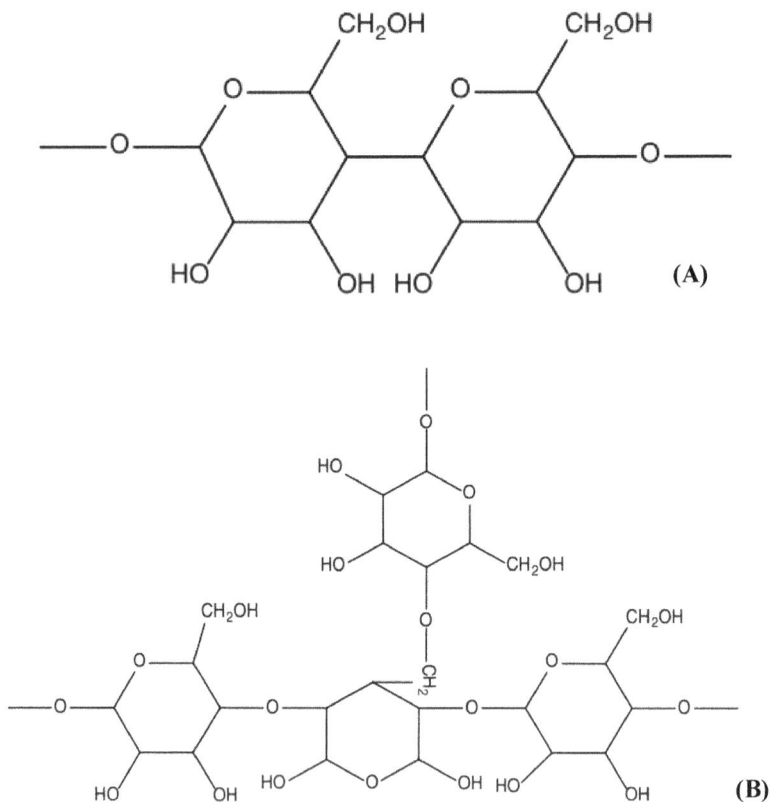

Figure 1.1 Structure of Starch granule showing (A) Amylose and (B) Amylopectin. Adapted from [14]

Chemically, amylose has a molecular weight of about 10^6 and a degree of polymerization that can be as high as 600. Amylopectin has a molecular weight in the region of 10^7 – 10^9, many glucose units that can be between 10 and 60, while not excluding side chains which accounts for the branched nature. Hydrophilicity of starch can be linked to the hydroxyl groups as can be seen in Fig. 1 above. It is pertinent to point out that the ratio of the two primary homopolymers influences physicochemical properties of starch whereas chemical modification, blending, pH etc. affects other properties such as mechanical and thermal properties [11]. One of the major treatment given to starch is gelatinization. Naturally starch occurs as granules which are insoluble in water but a process like gelatinization can be used to transforms aqueous starch solution into a paste increasing its wettability and dispersion in water. This is normally done by adding plasticizers such as glycerol which changes the structure of starch to a new form called thermoplastic starch. Gelatinized starch finds application in areas where mechanical strength and barrier properties is a priority for instance food packaging. Another form of starch is Carboxymethyl starch (CMS). This is synthesized by etherification of free hydroxyl groups of starch with carboxymethyl groups ($-CH_2COOH$). This derivative of starch is known for its mucoadhesivity and high pH responsiveness; features that are quite important for target drug delivery application [15]. Porous starch due to the presence of pores is another derivative of starch that has wide application in biomedical and pharmaceutical industry and have potential in waste water treatment.All in all, in the bid to continually improve the properties of starch various chemical modifications are carried out. These modifications are aimed at improving stability of starch pastes at high temperature, improving opacity, texture and freeze – thaw stability. It is pertinent to point out that most of these modifications depends on the reaction between the –OH groups of amylose and amylopectin [16]. These class of starch have lower gel formation rate, improved clarity and lower pasting viscosity than normal starch with application in producing confectioneries. Other kind of starch with areas of applications are dextrin (encapsulating agents and in emulsion), starch ethers like hydroxyethyl starch and hydroxypropyl starch (fruit pie fillings), starch esters like monostarch phosphates and distarch phosphates (fruit pie fillings), starch acetates, starch adipates and succinates (condensed soups).

Starch based composite like starch-clay composite have been seen by [17] to have potential in paper making industry. Addition of this composite as a filler in papers was observed to increase bonding, tensile, burst and folding strength as well as the optical properties of the paper. Also, starch-based scaffolds have been used for adhesion, proliferation, differentiation and regeneration of cells [18]

1.2   Sources of starch

Carbohydrate is stored in the form of starch by plants. These are produced when sun rays trapped by leaves of plants catalyzes the reaction between carbon dioxide and water with oxygen as the by-product. All such plants produce carbohydrate but not all store it in quantifiable manner as starch as to attract commercial interest.

Starch sources from plant can be obtained as categorized below.

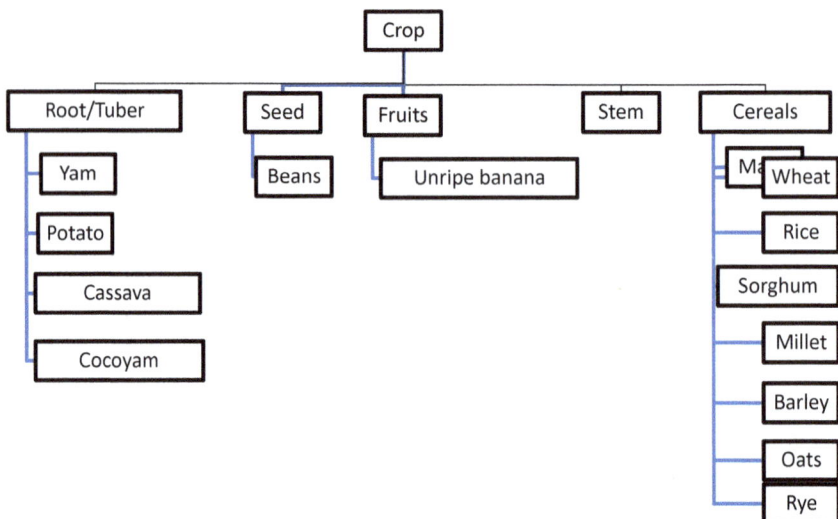

```
                              ┌──────┐
                              │ Crop │
                              └──────┘
                                  │
   ┌────────────┬──────────┬──────┴──────┬──────────┬──────────────┐
┌──────────┐ ┌──────┐ ┌────────┐     ┌──────┐   ┌─────────┐
│Root/Tuber│ │ Seed │ │ Fruits │     │ Stem │   │ Cereals │
└──────────┘ └──────┘ └────────┘     └──────┘   └─────────┘
   │           │          │                         │
 ┌─────┐    ┌──────┐  ┌─────────────┐            ┌──────────┐
 │ Yam │    │Beans │  │Unripe banana│            │Ma│ Wheat │
 └─────┘    └──────┘  └─────────────┘            └──────────┘
 ┌────────┐                                      ┌──────┐
 │ Potato │                                      │ Rice │
 └────────┘                                      └──────┘
 ┌─────────┐                                     ┌─────────┐
 │ Cassava │                                     │ Sorghum │
 └─────────┘                                     └─────────┘
 ┌─────────┐                                     ┌────────┐
 │ Cocoyam │                                     │ Millet │
 └─────────┘                                     └────────┘
                                                 ┌────────┐
                                                 │ Barley │
                                                 └────────┘
                                                 ┌──────┐
                                                 │ Oats │
                                                 └──────┘
                                                 ┌─────┐
                                                 │ Rye │
                                                 └─────┘
```

Root/tuber crops are among the most important source of starch. These class of crops are grown for their thick stems or roots which normally develop underground. Root and tuber crops are the stable food in the tropics as well as feed for animals. Root and tuber crops due to their high starch content generally contain low levels of protein, mineral and vitamin content. Cereals on the other hand store their starch in their grains. Compared to root and tuber crops, cereals have lower starch content. Plants also store their food (carbohydrate) in their seeds as seen in some leguminous plants such as beans, Bambara nut while others can be seen in their fruits such as plantain. All in all, the percent composition of starch found in each kind of plant is dependent on many factors like specie, variety and environment among others. As such, it is more convenient to analyze the starch composition of plants by the percentage composition of the two polymers that make up starch granule (amylose and amylopectin).

Below is a table showing the percentage composition of amylose and amylopectin of various plants.

Table 1. Starch, Amylose and Amylopectin content of various crops [12].

	% Amylose	% Amylopectin
Yam	22	78
Potato	21	79
Rice	17.5	82.5
Corn	24 – 28	75
Sweet Potato	18.9	81.1
Wheat	21.7	78.3
Barley	27.5	72.5
Waxy maize	<1	>99
Waxy Corn	0	100
Maize	21.5	78.5
Chestnut	19.6	80.4
Arrowroot	25.6	74.4
Edible Canna	22.2	77.8
Lentil	29 – 45	54 – 71
Lotus Root	15.9	84.1

1.3   Properties of Starch polymer composites

Starch and starch based composites though cheap are known for a series of interesting properties. Physically starch has a variable density depending on source, environment among others. Starch is a hydrophilic material in nature and this can influencethe density of the granules. However, Al-Muhtaseb et al. [19] observed that in potato starch and gel, bulk density of starch has an inverse relationship with moisture content. This was ascribed to the kind of interaction starch has with bound water and free water. Volumetric shrinkage also decreased with moisture content. Starch is white, formless and insoluble in chilled water but changes physically when placed in boiling water. Under a magnifying lens, it is seen in the form of spheres or granules. Starch unlike other biodegradable polymers retains its appearance till a critical temperature which is dependent on source of starch is attained. Starch is known to be brittle, difficult to extrude and as such requires ways to make it more paste-like [20]. Ability to gelatinize (form thick pastes on heating in presence of excess water and or a plasticizer like glycerol) is another distinguishing property of starch. During this process, an irreversible disruption of the molecular order with the starch granule occurs. This is normally accompanied by diffusion of water into the granules, radial swelling of granules, loss of crystallinity and leaching of amylose into the aqueous solution. Swelling of granules produces a viscous paste made of a continuous phase of solubilized amylose and/or amylopectin, and a discontinuous phase of granule fragments which on cooling, amylose and linear segments of amylopectin re-organize to form an ordered gel network, an occurrence often seen as the first stage of crystallization of starch molecule. Due to the weakened hydrogen bonding in the amorphous areas, gelatinization tends to start there with the onset temperature depending on the source of the starch [21]. Gelatinized starch on storage tends to become increasingly less soluble; an undesirable process called retrogradation [22]. Conversely retrograded starch though undesirable for food packaging application due to poor freeze – thaw stability is considered a dietary fiber.

Gelatinization of starch makes it more plastic leading to the formation of a derivative of starch called thermoplastic starch. Thermoplastic starch are quite water sensitive and thus can exhibit poor mechanical properties due to their hydrophilicity [23]. It is pertinent to point out that the temperature at which a particular starch gelatinizes differs and is an important factor in film forming application. For starches from root and tuber crops, gelatinization has been observed to occur at relatively low temperatures, with swift and even swelling of the granules. However, unlike cereal starches they retrograde while displaying high paste clarity and viscosity profile [24]. Abdullah et al. while studying the physical and chemical properties of potato, corn and cassava starch observed that cassava exhibited highest gelatinization temperature of 68 °C followed by corn (66 °C) and potato (62 °C). Gelatinization temperature was seen to be dependent on size of starch granule and degree of crystallinity which is also dependent on chain length of amylopectin. Morphological analysis showed that potato starch has largest granular size followed by corn and then cassava [25]. However unlike thermoplastic and natural starch, a starch derivative like carboxymethyl starch (CMS) is soluble in cold water and does not exhibit the undesirable tendency to retrograde (crystallize upon heating to separate into two constituents) [15].

Naturally starch has poor mechanical strength thus the need for reinforcing with fillers/fibers and or blending with other polymers.Incorporation of natural fibers like sisal [20], eggshell [26], graphene oxide [27] nanoclays [28]; starch nanocrystals [29] have led to improvement in mechanical, barrier and thermal properties. This is due to the formation of covalent bonds with the reinforcing phase. Egg shell for instance was observed to improve the mechanical and barrier properties of corn starch films due to induction effect of C-C group in cornstarch and O-C-O group of calcium carbonate in eggshells [26]. Vaezi et al. while exploring the influence of zinc oxide (ZnO) nanoparticle on properties of thermoplastic cationic starch/montmorillonite (MMT) films observed that incorporation of zinc oxide and MMT led to an increase in tensile strength, opacity and a decrease in elongation at break, light transmittance, moisture absorption, film solubility and water vapour permeability. The formation of hydroxyl bond between cationic starch, MMT and zinc oxide as confirmed by FT-IR (Fourier Transform Infrared spectroscopy) examination was the plausible reason given for these observed behaviors [28]. Naderizadeh et al. [29] also reported the enhancement of starch/PVA blend by the addition of starch nanocrystals and sodium montmorillonite (Na-MMT). Addition of Na-MMT improved the tensile strength and modulus and decreased the elongation at break which is in agreement with the result of [28]. The ability of nanoclays to form hydrogen bonding with the polymer chains, not excluding the effect of their aspect ratio was observed to be a factor that leads to increase in mechanical properties. Starch nanocrystals proved to be a more effective material in improving the mechanical properties when compared with Na-MMT. This can be linked to their higher surface area as a result of smaller particle size thus forming more hydrogen bonds. Thermal stability of the blend were also improved due to the addition of the nanomaterials [29]. Although introduction of these materials improves the properties of starch composites, starch content plays a major role. Luchese et al. [30] observed that mechanical properties like tensile strength, elongation and elastic modulus improved with increasing starch content for both cassava and corn starch films. Water vapour permeability and water solubility decreased with increasing starch content for both starch sources although corn starch showed lower solubility and higher permeability than cassava starch. This was attributed to higher interaction between glycerol and intermolecular forces existing between corn starch as compared to that of cassava starch. With increased starch content, increased film resistance and stiffness is guaranteed although filmogenic solution viscosity can interfere in the re-association of starch molecules thus affecting film properties as seen in the case of cassava starch.

Another important property of starch is it affinity for water. For application like food packaging, low solubility and water vapour permeability is desired, however hydrophilicity of starch has proved valuable in adsorbent usage as it helps to improve dispersion in a given aqueous solution [31]. Since most fillers incorporated in starch matrix are hydrophobic, starch composite tends to be less hydrophilic than natural starch. Coating have also been seen as a way to improve on this property. Soares et al. [32]

observed that coated blend of thermoplastic starch/Poly(lactic acid) sheets were less soluble in water and also possessed improved strength. Starch can be treated to have both surface and closed pores. Various kinds of pores exist based on their size micropores (<2 nm), mesopores (>2 nm < 50 nm) and macropores (>50 nm) [22]. In selection of pores for a biomedical application like producing scaffolds for tissue regeneration, pore size, pore dimension and pore geometry are critical factors. Thermally, starch has a low glass transition temperature (low thermal stability).

All in all, native starch due to its inherent properties which makes it to be sensitive to processing conditions like pH, temperature, shear rate, freeze – thaw variation among others have limited direct application. Nonetheless, it is used in the pharmaceutical industry in producing capsules and tablets [22].

1.4 Chemistry of starch composites

Starch is a biodegradable polymer with interesting chemistry which have been seen to be quite influential on the properties displayed. Generally, starch are made of granules with morphologies that varies depending on biological origin (biochemistry of chloroplast as well as physiology of the plant source) [21]. According to Mishra and Tai [33], starch granules can be polyhedral, spherical or flattened ellipsoidal as seen in the case of corn, potato and tapioca starch. An average granule size of 12.2, 30.5 and 15.0 μm was observed for corn, potato and tapioca starch respectively. This agrees with the observation of Luchese et al. [30] who observed a size of 12.2 and 11.9 μm for corn and cassava starch as well as a polyhedral shape for corn and rounded or oval shape for cassava starch. Teixeira et al. [34] observed an oval shaped granule with size ranging between 5 and 20 μm in native cassava starch. Starch from Tamarind kernels were seen to have an oval shape with granule size of between 2.9 – 6.15 and 2.9 – 6.01 μm [35]. It is pertinent to point out that starch granule size affects properties like crystallinity, pasting, solubility, and enzyme susceptibility although other factors like amylose content, tightness of starch granule among others cannot be ruled out. Starch from wild mango seed (*CordylaafricanaL.*) *as* studied by Ngobeseet. al. was seen to be truncated (cap-shaped) with a size of 5.7μm, which is much smaller than that of bean (28.2 μm), pea (23.0 μm) and maize [36]. Root tuber crop like arrowroot showed granule sizes ranging from 9.47 to 22.47μm with a mixture of large to small oval to elliptical granules [24].

Generally, the XRD diffractograms of starches show the crystalline amylopectin structure of starch. X-ray diffraction analysis of starch from different sources shows three common kinds of pattern namely A, B and C. A-type pattern is shown by mostly cereal starches whereas B-type pattern shows true crystalline form of starch. C-type pattern on the other hand looks more like an overlaid A and B-type patterns [37]. *Cordylaafricana L.,*a leguminous fruit tree common in Southern Africa exhibits a C-type diffraction pattern similar to that of bean and pea starch [36]. A-type pattern was observed of Tamarind (*Tamarindusindica L.*) [35]. A review of the molecular structure of root and tuber crops by [37] showed that most root and tuber starch have a B-type pattern with the exception of a few like *Ipomea batatas, Manhiot esculenta, Nelumbo nucifera, Dioscorea dumetorum and Rhizoma dioscorea*. Growth condition, biological origin, maturity, amylose/amylopectin content among others accounts for the differences in diffraction patterns observed. Viscosity of film forming starch suspension have been seen to depend on morphology, size and roughness of the starch source (morphology). Li et al. [38] while irradiating native corn starch with ^{60}Co gamma ray observed a change on the molecular structure of starch leading to a decrease in crystallinity. Improved tensile strength and a significant decrease in water permeability were observed. The ability of the rays to break up the crystalline region in starch granules can be the main factor. Fourier transform infra-red (FT-IR) analysis confirmed that polysaccharides like starch and chitosan interact through hydrogen bonding often shown by the shift in the observed peaks in FT-IR and reduction in crystallinity [39].

Summarily, changing the morphology of starch is the key to improving its properties. According to [40], beneficial properties of starch is dependent on its molecular structure although the details are yet to be fully understood.

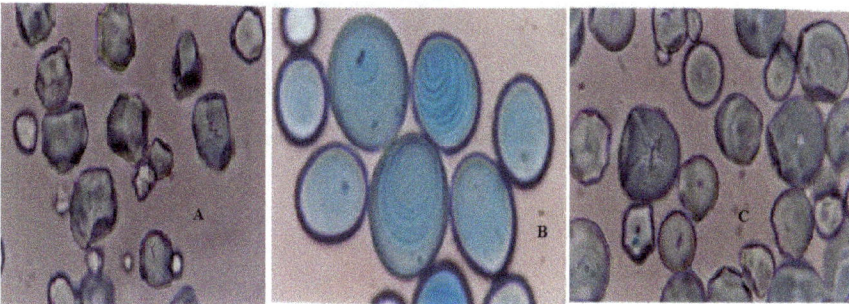

Figure 1.2 Optical Microscopy of (A) Corn Starch (B) Potato starch (C) Tapioca starch. Copied from [33].

Figure 1.3 SEM micrograph of (A) corn and (B) Cassava starch. Copied from [30]

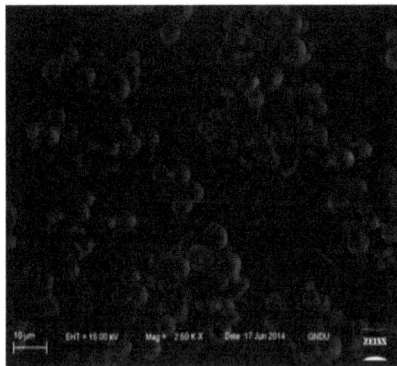

Figure 1.4 SEM micrograph of Tamarind (Tamarindusindica L.) Kernel Starch. Copied from [35]

2. Processing of starch composites

Processing is the major manufacturing step in the production of composites. Processing comes with conditions or parameters whose influence on the properties of the final product is quite significant. Similarly, the nature of raw material and the product to the formed also influences processing. Starch is a thermoplastic polymer though semi-crystalline in nature. Based on the kind of product to be manufactured, various processing methods have been devised and can be used. For instance, casting is more prevalent in the production of films for packaging application although one may look at it as a laboratory/small-scale/cheap method of producing films. Similarly, extrusion injection moulding and compression moulding are used when thicker parts are to be produced whereas extrusion cooking is used for foamed materials. All in all there are many kinds of processing of starch composites but a look at the chemistry of starch composites will give a further insight on how starch and its composites act during processing.

2.1   Processing techniques

Polymer composite processing generally involves a combination of both mechanical and thermal process. Whereas thermal input may be required for either melting the polymer of decreasing the viscosity, mechanical work is needed for proper dispersal and to ensure homogeneity. Although type of polymer, nature of filler or reinforcing material, size of the filler determines the processing technique, it is pertinent to state that the matrix and reinforcement need to remain intact at the end of the production run. Starch is a thermoplastic polymer and as such major processing techniques are extrusion, injection moulding, electrospinning, etc. A brief look at some of these processing techniques will be done below. However, the best technique for producing starch composites will depend on shape of part to be produced, cost and intended mechanical properties, optical and barrier properties [41].

Among the most significant application of starch and its composites, packaging is the most researched. Films for this operation are often produced by solution casting method. In solution casting, solvents are used to dissolve the polymers for instance water for starch, acetic acid, lactic acid, malic acid for chitosan [42,43] after which they are removed by drying. Although it is quite easy to perform and have low tooling requirement, it can be time, energy and labour intensive when the time for removal of solvent and drying of films are considered. As such, it is also difficult to scale up for batch processing [42]. Due to the nature of starch, extrusion is one of the most widely used especially for large scale production. Extrusion is a thermo-mechanical technique where the starch granules or blend is heated in a mandible and made to pass through a die under pressure. The properties of the produced composite depends on extrusion parameters like number of screws, temperature, pressure, L/D (length to diameter) ratio of the extrusion machine among others. Extrusion of starch with plasticizer involves granule expansion, gelatinization, decomposition, melting and crystallization [44].

Varieties of extrusion process include extrusion-cooking, reactive extrusion among others. Extrusion-cooking is an extrusion technique which entails extrusion of loose materials which are mostly natural raw (organic) materials under high temperature and pressure. This leads to development of significant physical and chemical changes in the processed material [45]. In the case of starch, extrusion cooking process destructures starch gelatinizing it thus making it easy to be molded and blended with other polymers. Extrusion-cooking is a derivate of the extrusion technology which is well-known in the plastic industry. It is commonly used in the production of food products like snacks, crispy bread, confectioneries, baby food among others. In extrusion cooking, pressure and existing shear forces are parameters that affect changes in the starch molecules. As such if the extrudates are to possess certain properties, setting the proper extrusion parameters are important. Use of screws with variable degrees of compression, die size, rpm are some of the ways [46]. It is pertinent to point out that pressure does not affect the viscosity of the extrudate as much as temperature does although it affects the viscosity and stability of the products especially those retained at a temperature of 95 °C.

Reactive extrusion on the other hand entails that the blending or reaction leading to the formation of the composite material take place in the extrusion chamber. It is often used in in-situ polymerization of starch composite or production of starch nanoparticles. In compression moulding, the material is laid between two moulds before heat and pressure is applied. It is also known as hot pressing. Compression moulding have the ability to produce large and complex shapes, design flexibility but tooling cost is high.Electrospinning is a fiber production method in which electrostatic force is used to draw out charged threads of polymer solution into fibers of nanosizes. Here ultrafine nanofibers are produced by charging polymer melt through a needle or spinneret under high voltage. It has been seen as an efficient method of producing nanofibrous biomaterials for biomedical applications. Orientation of fibers is dependent on the electrospinning setup. Electrospun fibers have been used by Lopez de Dicastillo et al. in dispersing cellulose nanocrystal in poly(lactic acid) with improved flexibility and ductility without a loss in barrier properties [47]. Poor mechanical strength, non-uniformity of pores are some shortfalls of the electrospinning technique.

Starch nanoparticles on the other hand can be prepared by methods like acid hydrolysis, enzymatic treatment, high pressure homogenization, ultrasonication, reactive extrusion, gamma irradiation among others [6,48]. An extensive review of processing techniques for starch based composite foams was carried out by [49].

2.3   Effect of processing techniques

Beside from many other factors such as nature of material, processing technique is one of the most prominent factors and affects not only properties of composite produced but the kind of product to be produced. Degree of adhesion and mixing has been seen to be dependent on the technique used. Whereas techniques such as hand lay-up is quite cheap and easy, it produces materials with lesser homogeneity when compared with processes such as injection moulding and extrusion. Mendes et al. blended chitosan with corn starch using extrusion method. Using thermoplastic chitosan of 5 and 10wt.%, thin films of about 800 μm were produced. The blended and pure polymer films were observed to showcase an amorphous spectra which is characteristic of semi-crystalline polymers. Corn starch was also observed to have been destructurized by the extrusion process. This was confirmed by morphological analysis where absence of starch granules were observed. However surface cracks were seen which can be attributed to presence of brittle chitosan. Presence of chitosan was seen to lead to a decrease in tensile strength and modulus but a better elongation and thermal stability [50]. Extrusion method was also explored by Toro-Márquez et al. for the production of active and intelligent materials for food packaging application. Using corn starch as matrix and pH sensitive nanofillers produced from natural and modified MMT nanopackaged with anthocyanins from *Hibiscus sabdariffa* (Jamaica), authors observed that exfoliation of nanofillers due to shear forces in the extruder led to the exposure of the anthocyanins leading to a decrease in expected properties [51].

Parameter like mixing time was explored by Amri et al. in the case of graphene oxide reinforced cassava starch. Increasing graphene oxide content and mixing time led to an improvement in tensile strength and modulus (3.92 MPa at 30% graphene oxide and mixing time of 60 minutes) whereas elongation decreased with mixing time which can be ascribed to improved homogeneity. However water uptake and biodegradation increased with graphene oxide content but decreased with mixing time. This can be ascribed to the hydrophilicity of starch and graphene oxide [27]. González-Seligra et al. reported the effect of screw speed on the properties of cassava starch film. At a speed of 80 rpm plasticized starch films showed a uniform rough texture unlike at 40 and 120 rpm where broken starch grains were evident on the film surface. Low screw speed can lower the specific mechanical energy needed to process the starch grains whereas very high screw speed may lead to short extrusion time thus causing some of the starch not to be processed. Presence of starch granules on the surface of films processed at 40 and 120 rpm led to a reduction in plastic zone. However starch films processed at 80 rpm exhibited lower tensile strength at break and modulus. Water solubility decreased with increasing screw speed whereas 40 rpm starch film showcased highest water vapour permeability although plasticizer concentration also plays a role in solubility [44].

Dang and Yoksan formulated a starch/chitosan blend using blown film extrusion method. This method though challenging promises continuity in production, low energy demand, and decreased production time. Presence of chitosan improved the water and oxygen barrier properties, and minified the hydrophilicity of the film which was ascribed to hydrogen bond interaction between starch and chitosan [42]. Brandelero et al. also explored the effect of production flow technique on the properties of starch/poly(butylene-adipate-co-terephthalate) (PBAT) films produced by blown extrusion technique. The films were produced by both extruding pellets of thermoplastic starch with PBAT pellets or extruding starch granules with PBAT and that of glycerol. PBAT concentration was seen to play a significant role in the properties of the films produced. The former method produced films with better properties when PBAT concentration was 50% whereas for concentration above 50%, the latter method proved to be more effective and at a lower cost [52].

Quiles-Carrillo et al. combined extrusion method with injection moulding in manufacturing of ternary blend of Polylactide, poly(ε-caprolactone) and thermoplastic starch. Mechanical test showed that ternary blends showcased higher ductility than the binary blends showing synergy between starch and poly(ε-caprolactone)[53]. Muller et al. [54] in a bid to produce thermoplastic starch films, reinforced thermoplastic starch with hydrophilic and hydrophobic nanoclays via extrusion and thermopressing technique. Thermoplastic starch was produced from cassava starch and glycerol whereas nanoclay used were organically modified montmorillonite (MMT) called Cloisite 30B and an unmodified MMT (Cloisite Na^+). Using a single-screw extruder with a 25 mm screw diameter and 4 heating zones, the mixture was extruded at a screw speed of 35 rpm before thermopressing at a temperature of 110 °C and a subsequent pressure from a 4 tons load. Hydrophilic nanoclay as expected led to a decrease in the water vapour permeability due to formation of an intercalated composite unlike the hydrophobic nanoclay which not only increased the tensile strength, reduced water permeability but formed an exfoliated composite. As such incorporation of nanoclays as a way of forming films offers a way to improve their properties for industrial application. Nevertheless, the level of dispersion depends more on the hygroscopicity and concentration rather than the combined processing method.

Ren et al. [39] studied the influence of chitosan concentration on the properties of corn starch/chitosan films produced by casting method. Increase in chitosan concentration led to an improvement in film solubility, color, tensile strength, elongation but a decrease in young modulus, water vapour permeability and crystallinity. Compared with the observations of [55], a lower tensile strength and higher elongation at break was recorded although higher concentration was used. As such external factors like relative humidity might have to be considered as can been seen in the values of relative humidity used. A differing trend of water vapor permeability was also observed when result obtained is compared to the observation of [42] but agrees with the result of [55]. Zakaria et al. characterized thermoplastic potato starch/bentonite nanocomposite film produced by solution casting technique. Using bentonite of 1, 5, 10, 15 and 20% weight fraction, tensile strength and elongation increased with bentonite content thus showcasing the reinforcing effect of bentonite [23]. Zakaria et al. [56] also explored the effect of mixing temperature on the properties of potato starch film produced by solution casting technique. 85 °C showed the best combination of tensile strength and elongation at break among the four temperatures considered (80, 85, 90 and 95 °C). High mixing temperature leads to structural modification of starch whereas low mixing temperature proves to be insufficient for complete gelatinization of starch thus leading to rough morphology with swollen starch granules.

Compression moulding technique was used by Naik et al. [20] to process sisal fiber reinforced starch composite while looking at the effect of fiber length on the mechanical properties of the bio-composite. Results showed that tensile, impact and flexural strength increased with fiber length. Composite with fiber of length 20mm showcased the highest strength. Da Róz et al. also explored the effect of molar mass on the thermal and mechanical behavior of a hot pressed and organic acid modified thermoplastic starch. The authors used ascorbic acid and citric acid to reactively process thermoplastic starch produced from corn starch and glycerol. Processed samples were hot pressed into 2 mm plates. Result showed that increase in concentration of the acids leads to a decrease in viscosity and an increase in tack stickiness. This was attributed to reduction in molar mass. Addition of the acids also led to production of clearer mixtures although samples containing ascorbic acid tend to change color on storage. X-ray spectra, crystallinity and DSC (Differential Scanning Calorimetry) results were not affected by the reaction with organic acids but melting temperature decreased. This was ascribed to decrease in size of crystallite and presence of impurities [57]. Thunwall et al. also used compression moulding method to produce starch sheets/films. Two classes of starch namely, high amylose potato starch and normal potato starch were used in producing the sheets. After conditioning at 23 °C and a relative humidity of 53%, high amylose potato starch exhibited higher tensile strength and modulus. Processing parameters such as temperature and moisture content were seen to have effect on the mechanical properties of normal potato starch [58]. Ortega-Toro et al. also processed starch/polycaprolactone bilayer film by compression moulding. One layer was made from polycaprolactone and thermoplastic starch whereas the other was made from polycaprolactone and thermoplastic starch with 5% polycaprolactone. Bilayered films showed improved mechanical and barrier properties [59]. Lopez et al. produced a thermoplastic corn starch films reinforced with fibrous residue from *Pachyrhizus ahipa* via melt mixing and compression moulding method. Lower water vapor permeability and improved tensile strength and modulus were observed. However introduction of the fibrous residue led to increased film roughness [60].

Bubble electrospinning was used by Liu and He [61] to fabricate a PVA/Starch nanofibers of varying ratios namely 10:0, 3:1, 1:1, 1:3 and 0:10. Blend ratio was observed to be an important factor in the processing of the blends. Below a starch/PVA ratio of 1:1, droplets were formed. However as the ratio of PVA increases, the blend becomes easier to process with smooth morphology and even diameter. Starch is not easy to process especially at low temperature but addition of PVA makes it more spinnable. Similarly rheological studies shows that shear thinning was not observed in the blends of 1:1 and 1:3 at low shear rate thus the even morphology and excellent spinnability exhibited. Wu et al. also compared direct mixing and co-coagulation technique in the processing of starch/rubber composite. Co-coagulated blends showed better mechanical properties than the direct mixed blend. This was attributed to improved dispersion offered by co-coagulation [62]. Wanter and Mangindaan precipitated Indonesian taro starch using sodium hydroxide, urea and ethanol. Precipitated starch nanoparticles were amorphous with diameters of about 203-596 μm [63]. Conversely, [29] prepared starch nanocrystals with average size of 230 μm and crystallinity of 57% from potato starch via acid hydrolysis.

Starch and its composites are also used in manufacturing of foamed materials; a widely used material in packaging of products to be transported. Combrzyński et al. explored the use of extrusion-cooking technique in producing foamed materials from starch while studying the effect of temperature, die type, humidity, screw speed, raw material type and additive blend. Using wheat, potato, corn and corn chemically modified starch, result showed that dextrinization of starch like potato occurs above 120 °C. Among all the additives, only Plastron foam PDE and poly(vinyl) alcohol PVA exhibited satisfactory result for production of foamed materials. Products with more than two blends had no pores, exhibits compact structure and an uneven shape. Similarly humidity level below 17% and above 19% had significant effect on the process. Lower die diameter gave best expansion ratio. Higher screw speed of (130 rpm) allowed for products with high expansion ratio to be produced. Most of the starch used produced foamed materials [45]. The influence of ultrasonication treatment on properties of corn starch was studied by Amini et. al. [64]. Temperature, exposure time were observed to influence the properties of the starch whereas ultrasound amplitude and concentration had insignificant effect. Gelatinisation enthalpy and temperature range decreased due to ultrasonication. A change in the pasting behavior of the starch samples were also observed. Ultrasonication can affect intermolecular bonding in the starch granules.

3.   Why blend polymers

Although biodegradable polymers and its composites are known for their amazing specific properties, varying shortcoming exists in areas like mechanical strength, thermal stability, cost of production among others. Polymers like polylactic acid (PLA) is known for its low thermal stability and high Tg which leads to exhibition of brittleness. Polybutylene succinate (PBS) being semi-crystalline exhibits better flexibility than PLA is appreciable. To overcome some of these shortcomings blending or mixing of polymers is done. According to Aigbodion et al. [41], blending is one of the simplest way to improve the properties of biodegradable polymers. Blending entails mixing of two or more polymers and as such birthing a new polymeric material with improved properties. Free energy and entropy of mixing dictates the extent of miscibility of the blended polymers. This can be represented by

$$dG_M = dH_M - TdS_M$$

Where G_M, H_M and S_M are free energy, enthalpy and entropy of mixing respectively for the polymer blend. Thermodynamically, a negative free energy shows that proper miscibility which can lead to improved properties can be achieved. However temperature is an important factor in thermodynamic processes.

3.1   Starch and starch blended polymers

One of the major challenges of biodegradable polymers like starch lies in their high affinity for water, increased moisture permeability among others. For healthy competition and proper replacement of synthetic polymers, new ways of improving both physical and chemical properties will aid in maximizing the potentiala of these polymers. Various techniques like physical modification which involves blending with other polymers and or reinforcing with particulates/fillers exists. Properties of physically modified polymers are dependent of factors like interfacial adhesion, phase morphology, volume fraction, etc. The chemical modification alters the structure of the polymers and helps to improve the cross-linking ability of the polymer. As such, various chemical modification techniques include cross-linking, acetylation, phosphorylation among others. A combination of physical and chemical modification can also be done. Modified starch has improved film forming ability, higher water solubility and improved potential for coating formation [65]. However mechanical properties of blended films depends on chain stiffness, molecular symmetry of the polymers and intermolecular forces [66].

Abdul Khalil et al. reported the effect of corn starch and red seaweed blend ratio on the physicochemical and mechanical properties of the developed film. The hydrophilic nature of starch and seaweed were observed to play important role in determining the barrier and mechanical properties. As the ratio of red seaweed to corn starch increased, an increase in thickness, tensile strength, water vapour permeability, young modulus with a nonlinear behavior for elongation was observed. Besides, κ-carrageenan, a major polysaccharide in red seaweed has a negative charge which allows for proper bonding with positively charged groups in starch [67]. Similar study by Garalde et al. [68] for starch/PBAT films revealed that increasing the ratio from 20:80 to 40:60 leads to improvement in the dispersion homogeneity of the blends, an increase in crystallization temperature of PBAT and a reduction in melting transition and tensile properties. The authors also observed that when the blend where left for 3 months, only the tensile strength, modulus and elongation of 60:40 blend ratio increased which was attributed to plasticizing effect caused by the migration of glycerol from the starch phase of the blend. Tensile strength and elongation of 20:80 blend ratio decreased whereas elongation of 40:60 ratio remained roughly the same. Thermal properties were also affected by the storage time. Katerinopoulou et al. reinforced a blend of Poly-vinyl-alcohol (PVOH) and maize starch with nanoclays using glycerol as plasticizer. Nanocomposite containing 10wt% glycerol and 20wt% PVOH showed highest strength and stiffness while exhibiting appreciable brittleness. Composites made from 20wt% glycerol and 10wt% PVOH showcased lower strength and stiffness but have superior elongation and film forming capacity. The increase in brittleness was attributed to increased formation of hydrogen bond between PVOH and starch. Incorporation of both PVOH and nanoclay in the maize starch improved its thermal stability by 50^0C but led to about 74% reduction in water vapour transmission rate when compared to plasticized maize starch film. Conversely, biodegradation of the films increased with increase in starch content unlike when PVOH was increased. Based on observed properties nanocomposite containing 10wt% glycerol and 20wt% PVOH will find application in durable pot packages whereas 20wt% glycerol and 10wt% PVOH will make a good film material for flexible film packaging application [69]. Bonilla et al. [55] also explored the effect of chitosan addition on the properties of wheat films. Increasing the ratio of chitosan in the starch blend improved the mechanical and barrier properties of the film. The films when used to wrap pork meat showed antimicrobial activities which also depends on the ratio of chitosan and agreed with the observation of [43] on the potential antimicrobial properties of starch-chitosan blend.

One of the ways of modifying starch composites is by crosslinking using chemicals like glutaraldehyde. Li et al. explored the effect of crosslinking on the properties of starch-chitosan blend. Crosslinking with glutaraldehyde was seen to lead to deterioration of the

Advanced Applications of Polysaccharides and their Composites   Materials Research Forum LLC
Materials Research Foundations **73** (2020) 198-241   https://doi.org/10.21741/9781644900772-7

mechanical properties, insignificant change in moisture uptake and water vapor transmission rate with a decrease in thickness swelling. Addition of glutaraldehyde leads to embrittlement of the films by damaging the ordered structure of the films thus limiting the improvement chitosan adds to starch [66]. The crosslinking ability of glutataldehyde and citric acid on starch/polyvinyl alcohol blend was compared by More et al. Compared to glutataldehyde, citric acid improved the flexibility, yellowness index (transparency) and water vapour permeability of the films but showed a minimal decrease in tensile strength; 3.53MPa as against 4.83MPa for glutaraldehyde crosslinked films. Thus citric acid seems to have potential as a crosslinking agent for starch blend with potential application in food packaging industry [70]. Kumar et al. [71] also found out that increasing the concentration of citric acid as a crosslinking agent leads to an increase in thickness, tensile strength but a decrease in moisture content, opacity, swelling index, elongation, solubility and water vapour permeability. The ability of citric acid to form covalent bonds with starch helps to reduce the free hydroxyl group in the starch matrix thus reducing the water holding capacity or hydrophilicity of starch [70, 71]. Citric acid can also reduce chain mobility which may affect flexibility. Garcia and Turbiani explored the use of sericin as compatibilizer for starch-PBAT films. Introduction of sericin leads to an improvement in tensile strength and modulus whereas elongation and water vapour permeability reduced thus highlighting the ability of sericin to reduce interfacial tension and enhance compatibility [73]. Effect of varying chemical modification process on the properties of starch from native sources was carried out by Sukhija et al. The authors chemically modified native lotus rhizome starch using sodium hypochloride and varying concentration of chlorine (oxidation). Cross linking was carried out using varying concentration of sodium trimetaphosphate (STMP). Dual modification was done by first crosslinking with sodium trimetaphosphate (STMP) before oxidizing using varying concentration of active chlorine. Native and modified starch were blended with whey protein, psyllium husk and glycerol and used to produce films. Thickness measurement showed that modified starch had higher thickness than the native starch with oxidized starch composite film showcasing highest thickness. Increased interaction between glycerol and carboxyl/carbonyl group led to water retention during drying of the films and increased thickness. Improved lightness was observed in the oxidized starch due to oxidation of pigments present in the starch as compared the other two modification method. Presence of phosphate groups in the modified starch caused the starch to be more hydrophilic unlike native starch. Oxidized starch showed highest moisture content. Moisture content was also seen to increase with oxidation level. Dual modified starch composite exhibited lowest solubility and water vapour permeability when compared to oxidized and crosslinked starch composite films because cross-linking before oxidation improved the structural integrity of starch. Dual modified starch films also showed highest tensile strength and elongation due to the presence of phosphate groups that led to improved inter- and intramolecular linkages and enhanced molecular interactions. Cross-linked films offered the highest opacity whereas oxidized starch films showed best transmittance within the UV (ultra-violet) region. Due to its transparency oxidized starch composite film can be used where visual inspection of packaged material is of utmost importance [74].

Biopolymers are known for their brittleness and fragility thus making them difficult to form via thermal processes. Since thermoforming techniques are the most widely used fabrication techniques, improving the workability of the polymers is essential. Plasticizers which are low molecular weight and non-volatile compounds plays an important role in this regard. Plasticizers improves flexibility, and ease of processing of polymers. They decrease the tension of deformation, viscosity, density etc. while increasing the chain flexibility, resistance to fracture and dielectric constant [75,76]. Garcia et al. reinforced cassava starch with waxy maize starch nanocrystals using glycerol as plasticizer. A 40% decrease in water permeability and a 380% increase in storage modulus at 50 °C was observed. Thermal analysis via TGA showed the formation of hydrogen bonding between glycerol and nanocrystals which led to improved filler/matrix interaction [5]. Thermal behavior of thermoplastic starch plasticized with 1-ethyl-3-methylimidazolium acetate and 1-ethyl-3-methylimidazolium chloride was studied by Ismail et al. [77]. Starch plasticized with 1-ethyl-3-methylimidazolium acetate was observed to promote low thermal degradation, exhibit higher peak degradation which was seen to be due to formation of stronger hydrogen bonding with thermoplastic starch , hence higher degradation rate compared to starch plasticized with 1-ethyl-3-methylimidazolium chloride. One of the most commonly used plasticizer is glycerol. Zakaria et al. explored the effect of glycerol concentration on the properties of potato starch films. Increasing glycerol concentration leads to a decrease in tensile strength whereas elongation at break increases till 30% glycerol. Anti-plasticization was observed to occur at high concentration (50 and 70%). Physicochemical analysis showed that thickness and moisture content of the films improved as the concentration of glycerol increased which can be attributed to hygroscopic nature of glycerol [78]. The influence of bioethanol as a plasticizer for stretchable starch films was also analyzed by Ekielski et al. [79]. The authors found out that alcohol (bioethanol) improves the Young's modulus and resistance to punch of the films as well as making the films less soluble in water.

Kumar et al. [71] blended potato starch with gelatin using sorbitol as plasticizer and citric acid as a modifier. Film thickness, moisture content, swelling index, water vapor barrier property, solubility, tensile strength, elongation, biodegradability, moisture absorption and optical properties were tested. Citric acid was observed to have no significant influence on the thickness of the composite films although a slight increase in thickness occurred as the concentration of the acid increased. This was attributed to reordering of the starch chains by citric acid. Conversely, the moisture content and swelling index of the films decreased with the percentage of citric acid. Cross-linking of the functional groups of citric acid and starch led to unavailability of free hydroxyl group in the matrix. Opacity of the films decreased because citric acid made the starch to be more crystalline and compact. Observed decrease in solubility and water vapor penetration was attributed to cross-linking and hydrogen bonding among the functional groups. Mechanical properties improved due to introduction of citric acid which was seen to aid crosslinking of starch chains. Mohd Zain et al. also used citric acid as a modifier for thermoplastic starch. Addition of citric acid was seen to improve the tensile strength, modulus and elongation at break with optimal combination of properties seen in composite modified with 2% citric acid. This was credited to the ability of citric acid in increasing the slippage of starch molecule. Citric acid leads to complete plasticization of starch as shown by the smooth fractured surface. Similarly, citric acid improves the bond between glycerol and water to starch thus leading to higher onset temperature and higher thermal stability. However as the concentration of citric acid increases, acidolysis can take place leading to the loss of rigid starch structure [80].

Wang et al. studied the thermal and thermomechanical behavior of a blend of a biomedical polymer like polycaprolactone and starch. Ratio of starch, Polycaprolactone and plasticizer was kept at 27:63:10. Presence of natural plasticizer was seen to decrease the melting and glass transition temperature of the polymer blend. However storage modulus of the starch blended polymer was seen to be higher than the pure polymer thus indicating the reinforcing effect of starch. Good damping behavior was observed in both the blend and pure polymer which can be important in orthopedic application [81]. Askari et al. [65] carried out physicochemical and structural analysis of psyllium gum and modified starch composite for film production. Varying weight fraction of modified starch and psyllium gum were blended together. Thickness, moisture content, solubility, water vapour permeability, tensile strength increased as the quantity of the modified starch increases. Increased thickness was attributed to gelatinization whereas ability of starch to create hydrogen bonds with water and decrease in compact structure of psyllium gum led to the observed behavior for moisture content and water solubility respectively. However compared with the work of [74], the composite films exhibited higher tensile strength. Thermal analysis showed that increase in the proportion of psyllium gum led to a reduction in thermal stability. Morphological analysis using SEM shows compactness and interaction between the polymers. Composite film in the ratio of 75:25 for psyllium gum and starch was seen to be best suited for food packaging application.

3.2   Challenges in the fabrication of starch and starch blended composites

Major challenge of starch and its composite lies in the area of structural strength and hydrophilicity.Starch is brittle with poor mechanical strength thus limiting its application. Though many fabrication techniques exist, most production process of starch based polymers take place under non-isothermal conditions which modifies the morphology and as such the properties of the formed polymers [81]. Complex chain conformation of polysaccharides like starch coupled with exhibited hydrodynamic responses and repulsive forces while in solution, negatively influence the spinning efficiency and production of reproducible nanofibers from this class of polymers [18]. Propensity to retrograde, moisture sensitivity among others have been seen to affect the electrospinning of starch. Influence of plasticizers also plays a role in the mechanical and barriers properties of starch blends [32].

The inability to guarantee the similar quality in different batches of materials is a general problem facing renewable and biodegradable materials as climate, geography, mode of collection, storage time among others affect physico-chemical and mechanical properties [14]. Similarly, processing techniques have been observed to impact structural changes on temperature and stress sensitive materials like starch. High temperature and shear forces encountered during a processing technique like extrusion leads to the degradation of starch. Moisture sensitivity is another challenge as some of the available processing techniques entail the use of a certain level of relative humidity of values which can be influenced by other processing parameters.

All in all, the path towards the full utilization of the potentials of starch and its composite blends looks tricky but with increased understanding of how the structure of starch granules influence its properties, the future looks bright for full replacement of synthetic polymer materials with green composites.

4.   Application

Starch based composites have generated a wide range of interest due to their advantages as matrix material. Environmentally, starch belongs to the class of materials that have no adverse effect on the environment. Cost-wise, use of starch is cheaper compared to most other biodegradable materials like cellulose. Microstructurally, starch can be modified and blended with other polymers, giving rise to an improved composite material. Modern society have placed more demand on the materials for food packaging application thus decreasing the role of older materials such as glass, metals, polyolefin plastics, etc. and introducing biodegradable polymer composites such as starch. Cost have placed an increasing demand for new materials or blend of materials for applications in areas like tissue engineering, biomedical and genetics. Most synthetic polymers such as are known to be costly thus limiting their use in these areas.

4.1   Food packaging

Being a perishable product, food quality and safety is crucial to solving the world food crisis. According to the Food and Agriculture Organization (FAO), about one – third of food produced global is wasted with majority of the wastage occurring during supply and due to improper storage. Decaying food is also a major contributor of methane gas, a major greenhouse gas. Comparing it with carbon emissions of other countries, carbon emission from food waste would rank as the third highest. Based on the increasing world hunger, it is imperative to find a lasting solution to not only food waste but factors that lead to food wastes. Exposure to moisture, microorganism, gases, mechanical shock, chemicals among others have been identified to lead to food spoilage. Proper packaging provides an avenue to prevent this. In addition, packaging also informs customers of a particular product. Mechanical properties, optical properties, vapour and gas barrier, aroma barrier, thermal properties and eco-friendliness are some of the attributes of a modern day food packaging material [82]. Modification and blending of starch composite offers a way of improving inherent properties for various applications. Kumar et al [71] studied the influence of citric acid on the properties of starch/gelatin composite with prospect of a potential material for food packaging application.Lin et al. studied the influence of coating packaging papers with starch/PVA films on moisture resistance of the paper. The authors observed that a coating weight of $4.0g/m^2$ is enough to improve the moisture resistance of the packaging paper. Introduction of a moisture resistant agent like glyoxal further improved the moisture resistance of the films [83]. Another important property of polymer films for food packaging application is the film thickness. Mechanical properties of the film as well as the moisture permeability of the films is dependent on film thickness. Opacity is another important property that is of prime importance because of aesthetics and appearance. For proper food reliability, solubility to water should be kept at a minimal. Starch is often used to improve the degradability of synthetic polymers. Liza et al. studied the degradation of linear low density polyethylene – starch – clay nanocomposite film under sunlight. Degradation was observed to start at starch granules [84].

Besides having good physicochemical and mechanical properties, films for food packaging application are also meant to possess antimicrobial properties which will help extend shell life of foods. Zhong et al. explored the antimicrobial properties of kudzu starch- chitosan composite films while also looking at the effect of acid solvent on the properties of the films. Films showed lower water adsorption ability, high water barrier property and antimicrobial activity against *Escherichia coli* and *Staphylococcus aureus.* The film having malic acid as solvent showcased the best result. Film having acetic acid showed best mechanical property, least solubility and lightest yellowness whereas films made from lactic acid showcased best flexibility [43]. Moreno et al. [85] found out that incorporation of antimicrobial/antioxidant protein like lyophilized bovine lactoferrin (LF) and lysozyme (LZ) in potato starch films changed the structural and physical properties of the starch films, influenced their thermal behavior and increased the glass transition temperature. Certain level of compatibility with starch chains were displayed by the introduced proteins via bond formations leading to the increase in transition temperature although some part separated and migrated to the surface of the films. Lactoferrin for instance increased the film's brittleness irrespective of water content. Antimicrobial action against *Escherichia coli* and *coliforms* and antioxidant properties were observed, thus showing potential for meat packaging application.

Solubility is very important because at high solubility, polymers are more likely to degrade. The transparency (opacity) of starch films depends on the assembling pattern of starch molecules in the films and is influenced by the presence of ordered zones which tend to decrease absorbance of light hence, increase transparency of the films [74].

4.2   Biomedical application

Biodegradable polymer materials have also shown potential as bone fillers or fixes of fracture. With appreciable mechanical strength and modulus, biodegradability and good interaction with body tissues, starch and its derivatives have emerged as a major material for various biomedical application.Mendes et al. noted that starch based blends and its derivatives such as starch/ethylene vinyl alcohol and starch/ethylene vinyl alcohol - hydroxyapatite are future materials for bone replacement [86]. The ability to mimic the activities of matrices that supports cells in their natural ambiance is one property a potential material for making scaffolds should possess. Porous nature of composites such as Bacterial cellulose/potato starch composite can be exploited in in-vitro application [9]. Use of scaffolds in tissue engineering of cartilages are crucial in tissue engineering as they support cell proliferation and determine the shape of the tissue [10]. Waghmare et al. experimented the use of starch nanfibres as scaffolds materials for wound healing applications using polyvinyl alcohol (PVOH) as plasticizer via electrospinning method. Starch to PVOH ratio of 30:70% w/w, a flow rate of 0.5 mL/h, voltage of 25 kV and tip to collector distance of 13 cm were seen to be optimal conditions for electrospinning of the blend. Crosslinking agents such as glutaraldehyde improved the mechanical and thermal properties although at high concentration, may be toxic to cells. All in all, pores in the blend provided good site for cell growth and proliferation making it a potential material for wound dressing [18].

Pore size, pore geometry and dimension have been identified as important governing factors in choosing materials for making scaffolds. This is because of the pores not only serve as medium of providing oxygen and nutrient to growing cells but as a place where cells multiply [18].

4.3 Drug delivery

The definitive purpose of any drug delivery system is to develop formulations for treating diseases. With the stiff resistance and long procedure in introducing new products in the pharmaceutical industry, the need to use low cost and biodegradable drug delivery agent becomes expedient. Although the growth of nanodrug delivery system is on the rise, shortcomings like increased toxicity which is as a result of increased surface area of nanoparticles have not been fully solved. As such use of nontoxic, cheap and environmentally friendly material like starch becomes a way out.Consequently, the use of biodegradable microparticles for a delivery agent is gaining attention with starch being one of the promising biodegradable polymers. It has been used for nasal delivery of drugs as well as delivery of orally or intramuscularly administered vaccines. Other areas include as a binder, diluent and disintegrant. This is due to the physicochemical properties of starch. Thus the perceived structural shortcomings of starch is a huge advantage in pharmaceutical industry. Pregelatinized maize starch for instance is used as a dry binder in tablets [87]. Pregelatinized starch characterized with its swelling ability is used as a hydrophilic excipient [87,88]. Pregelatinization involves gelatinizing and drying of starch suspension. Rodrigues and Emeje [88] in their review of the applications of starch and its derivatives in drug delivery outlined that starch is currently a sort after material in nanodrug delivery although most of the starch used were obtained from corn, rice and potato which can be ascribed to the influence of botanical source on the physicochemical properties of starch. Najafi et al. [89] reported the use of acetylated starch nanoparticles as a drug carrier for ciprofloxacin. The nanoparticles were seen to have high encapsulation efficiency. Calinesu et al. also used carboxymethyl starch with chitosan as a colon drug delivery agent [90]. Quadrado and Fajardo employed polyelectrolyte complexed carboxymethyl starch/ chitosan mucroparticles for gastrointestinal tract drug delivery and found out that microparticles have a pH dependent release profile which can help counter burst effect and non-sustained release [15]. Xiao et. al. [91] employed acetylated starch nanocrystals (ASN) from broken rice as a drug release agent for an anti-tumor drug such as doxorubicin hydrochloride (DOX). ASN exhibited a stable release rate, higher loading efficiency and loading content compared to starch nanocrystal. Acetylation of the starch nanocrystal might have led to a decrease in the rapid enzymatic degradation rate.

Among starch based carrier materials, porous starch is one that has attracted much attention due to its excellent adsorption property aided by the number of pores and hollows it possess [92,93]. Wu et al. formulated a biodegradable porous starch foam for improved oral delivery of lovastatin using solvent exchange method. When compared with commercial capsule and crystalline lovastatin, encapsulating lovastatin in the starch foam allowed for immediate release and improved dissolution which can be attributed to the ease of degradation of the starch foam by enzymes [94]. Similarly, Zhang et al. employed porous starch as a nano-delivery system for lipophilic probucol. Embedding probucol in porous starch increased the solubility of the drug, the release rate as well as its bioavailability when compared to the free drug suspension thus showcasing the efficacy of using starch as a drug delivery system. Li et al.optimized the processing parameters for the production of porous starch for use as a carrier to improve the oral bioavailability of melatonin (MLT); a poorly soluble drug. An optimal drug loading of 12.30% and encapsulating efficiency of 70.17% was obtained at a stirring time of 30 minutes, MLT concentration of 120mg/ml and a ratio of 1:5 for MLT and porous starch. MLT were seen to occupy the pores of porous starch attesting to the porous nature of starch. Bioavailability evaluation using rats showed that a maximum concentration of MLT and MLT-porous loaded starch of 134.26 and 291.77ng/ml were attained at 15 and 20 minutes respectively [92]. Santander-Ortega et al. formulated a propyl-starch based composite for use as a transdermal drug delivery agent. Using a simple emulsion technique that employs the use ethyl acetate rather, the derivate was observed to possess high encapsulating efficiency for flufenamic acid, testosterone and caffeine and a linear release rate that was attributed to the hydrophobic nature of the drug and polymer [95]. Beside drug delivery, porous starch has also find application in flavor delivery. Belingheri et al. [96] showed that porous starch can serve as a tomato flavor delivery agent. Solvent used to disperse the flavor on the starch was seen to be a primary factor affecting properties like shell life.

The hydrophilic nature of starch in addition to the spongy nature of porous starch has been a key property making it applicable as an adsorbent and even protective material for sensitive elements like food pigments, minerals, oils among others. However the nature of the pores is a prime factor determining the extent of its application. Various factors have been seen to affect this with the method of production being prominent although the origin of the starch granule cannot be left out. Benavent-Gil and Rosell [93] explored the effect of starch source and enzymes on the morphological and physicochemical properties of porous starch. Using amylolytic enzymes like Amyloglucosidase, α-amylase and cyclodextrin-glycosyltransferase on wheat, rice, potato and cassava, tuber starches were observed to be more resistant to enzymatic hydrolysis when compared to cereal starches. This was ascribed to high number of branch points in non-crystalline regions which led to high density amorphous regions and stable crystallites. Size distribution was seen to be dependent on starch source and enzyme. However starch source played more prominent role on the water adsorptive capacity of starches with cereal starch showing higher values with wheat starch treated with Amyloglucosidase having the highest value. This was attributed to the pore surface area.

4.5   Adsorption

Effluents from various industrial processes have been seen to be toxic to the environment. Conventionally, most of these effluents end up either on land or in water bodies and have been observed to cause serious harm to the for a and fauna inhabiting the environment due to the slow rate of degradation. Besides, the composition of effluent such as dyes have been seen to be carcinogenic. Although various techniques exist, adsorption is one of the most attractive due to simplicity of design, cost, ease of operation and efficiency [97]. With the increasing demand for production of low cost adsorbents, use of renewable, abundant and biodegradable materials such as starch comes to mind. As suchs the potentials of starch and its composites can be said to be far reaching. The non-toxicity, low cost, hydrophilicity and environmental friendly nature of starch and its composites makes it a potential material for adsorption of materials like metals and dyes. Chang et al. studied the characteristics of magnetic soluble starch functionalized with multiwall carbon nanotubes (MWCNTs) with potential use as adsorbents. Presence of starch is expected to improve the hydrophilicity and biocompatibility of the composite while acting as a platform for the growth of iron nanoparticles. Result showed that the MWCNT-starch-iron oxide composite exhibited superparamagnetic properties and was also a better adsorbent for anionic methyl orange and cationic methylene blue when compared with MWCNT-iron oxide. This was attributed to the hydrophilic nature of starch which helped to improve the dispersion of the composite in the aqueous solution [31]. Carbon nanotubes (CNTs) is one of the commonly used secondary material in starch based composite for various applications including adsorption studies. This can be linked to the difficulty in stabilizing CNTs in aqueous solutions to which incorporation in starch seems to solve. Yan et al. fabricated a starch/carboxylated multiwall carbon nanotube composite and observed the formation of –OH bonds between starch and CNT which also aided the dispersion of the composite in both water and chitosan films [98]. Ability to form stable solution without aggregation is imperative for good adsorption behavior. Zubair et al. [99] experimented the influence of different parameters on the adsorption behavior of starch –NiFe-layered double hydroxide composite. Layered double hydroxide generally known as anionic clays have received lots of interest in areas like water treatment and catalysis however incorporation in starch matrix helps to improve its eco-friendliness. The composite were prepared using co-precipitation method in the ratio of 1:1 and 2:1 for starch and NiFe-LDH.pH was seen as a controlling parameter affecting dye adsorption as it affects the surface charge of adsorbents, the ionization degree of different contaminants, the dissociation of functional groups on the active sites of the adsorbent, and the structure of the dye molecule. The percentage removal of dye methyl orange was seen to decrease with increase in pH from 3 to 4 with the composites showing a minor decrease at pH range of 4 – 7. This was ascribed to electrostatic interaction between the dye and the adsorbent due to protonation of the adsorbents as a result of low pH. However at higher pH, the number of positively charged sites reduces whereas negatively charged sites increases thus waning the interaction between the adsorbent and dye. Concentration of adsorbent was also seen to play a role in the adsorption behavior of the composite. Increasing the concentration of the adsorbent increases the quantity of dye removed with starch/NiFe-LDH composite in the ratio of 1:1 showcasing best result. Increase in the number of adsorption sites occurs with increase in quantity of the adsorbents. Similarly hydrophilicity and proper intercalation of starch in NiFe-LDH helps to improve adsorption. Isotherm study proposed that adsorption of the dye can be best described by Redlich Peterson isotherm model with adsorption capacity of 246.91 mg/g, 359.42 mg/g and 387.59 mg/g for NiFe-LDH, starch/NiFe-LDH (1:1), starch/NiFe-LDH (2:1) composites respectively. Thermodynamic study shows that increase in temperature leads to a decrease in adsorption of methyl orange thus showing exothermic behavior with adsorption mechanism being ruled by hydrogen bonding and ion exchange between the dye and the starch composite. A simple method of fabricating adsorbent was used by Gong et al. [97] in producing a magnetic carboxymethyl starch/poly(vinyl alcohol)composite gel for possible use as an adsorbent for removing catonicmethylene blue dye. Presence of carbomethyl starch was seen to improve electrostatic attraction between the composite gel and the dye. Adsorption kinetics of the composite gel was seen to follow a pseudo-second-order kinetic model which is similar to the observation of [99]. However pH was seen to have insignificant effect on the adsorption capacity [97]. Yu et al. [100] studied the cadmium ion adsorption capability of hydrogel consisting of laponite, starch and polyvinyl alcohol (PVA) processed via freezing/thawing process. Starch used in producing this nanocomposite is corn starch. FT-IR analysis shows that the cross-linking reaction between the polymers and laponite must have led to the formation of a new structure via hydrogen bonding. Microstructural analysis of the nanocomposite samples using scanning electron microscope showed that the manufacturing process (freezing/thawing) led to physical synergy between the polymers which on addition of the nanoclay, formed a more porous, uniform and stable network. Although melting temperature decreased, thermal stability increased as the quantity of laponite increased. 10% laponite gave the highest value of reswelling ratio whereas 12.5% laponite was seen to adsorb most Cd^{2+} ions after 24 hours and thus showing an increase in adsorption capacity as laponite increased. As such laponite acted and thus can be used as both a filler material and cross-linker for starch/PVA hydrogel.

Beside its use as an adsorbent, starch based materials have also shown potential as flocculants for wastewater treatment. Wang et al. developed starch-*graft*-poly(2-methacryloyoxyethyl) trimethyl ammonium chloride, a flocculant by grafting (2-methacryloyloxyethyl) trimethyl ammoniumchloride (DMC) on starch using potassium persulphate as initiator. The flocculant exhibited high flocculation capability for 0.25wt% kaolin suspension under acidic, basic and neutral condition. It also proved to be a good sludge conditioner. Starch made the copolymer to be more soluble and thus convenient for wastewater treatment [101].

5.   Future trend

Starch and its composites have been seen to showcase amazing potentials. With it being heavily abundant as compared to some other biodegradable matrix materials, one can say that it has a huge role to play in the future of ecofriendly composites. Although its deficits are explicit, more improvements can be obtained by proper understanding of some factors that influence its properties. The properties of natural starch is dependent on biological origin with the ratio of amylose and amylopectin being a major determining factor as they dictate the structure of starch granules. Proper understanding of this will help in tailoring the properties of starch and its composites besides solving one of the fundamental issues facing renewable composite materials in the form of uniformity of achieved properties. Incorporation of nanomaterials especially biodegradable ones have been seen as a credible way of enhancing the applicability of starch and its composites. However, more studies is needed on the migratory behavior of these nanomaterials especially when the intended area of application is in food or health related areas. This is of much essence due to the porous nature of starch; a property which have been exploited in drug delivery application as well as in adsorbents. Distribution as well as alignment of these nanomaterials can also be looked at as well as how the microstructure compares with observed properties. Similarly, the retro degradation and swelling behavior of nano-reinforced and natural fiber reinforced starch composites also requires more studies. Mechanical properties can be improved by addition of these materials but their polar nature can be having a poor influence on the water adsorption behavior. Coating of the reinforcement can be looked at.

Variety of available processing techniques is one advantage possessed by starch. Affixed with other factors such as varying relative humidity, quantity of plasticizers etc. a comparative analysis between similar works becomes difficult. High plasticizer can lead to increased solubility and affects the crystallinity of the matrix thus affecting the physico-chemical and mechanical properties. Extensive studies on optimization of these processing parameters in conjunction with processing techniques will be quite useful in future development of starch. Studies on the use of newer and natural plasticizers can also be looked at. Blending with other polymers have also shown advantages but phase morphology as well as encountered immiscibility affects the properties of the formed material thus more attention is required. Although crosslinking agents like glutaraldehyde and glycerol have been used to improve themiscibility, effect of storage time and condition can be looked at as a possible migration of one of the phases with time cannot be ruled out.

All in all, the future of starch composites as a major green composite material seems bright. Introduction of advanced manufacturing technique such as fused deposition modelling, direct energy deposition will also be beneficial in advancing the course of starch composites as well as solving some of the aforementioned issues.

Conclusion

Environmental policies coupled with increasing concern over waste management system have pushed for a worldwide interest in the development of eco-friendly materials as well as proper utilization of wastes. Among various renewable, low cost, environmentally benign and easily available polymers, starch stands out due to some daring physico-chemical properties as well as ease of accessibility. Besides its properties which have helped it in finding applications in diverse areas, its inherent shortfalls are also evident. Starch is brittle, hydrophilic with poor mechanical strength and retrogrades. Expanding the allied areas of application will now depend on circumventing these shortfalls. Incorporation of nanosized biodegradable materials have been seen to offer a possible way out with improved properties obtained at low per unit cost. Excellent miscibility and good interfacial bonding is expected between starch and these materials since they are polar. Treatment of starch (plasticization) has also helped to improve its barrier and water adsorption properties. Blending with other biodegradable polymers have also been seen to improve its mechanical properties as have been shown by many research works reviewed in this paper. Thus one can say that the future of starch is huge as more areas of application continues to open up.

However, more work on improving the microstructure, retro-gradation and swelling behavior of starch is needed. Shell life of starch composites requires more attention as exposure to different environment (varying moisture, pH etc.) especially in application like food packaging will affect its performance. Integration of advanced manufacturing techniques will also help improve future use and lower cost of starch composites.

References

[1] C.C. Daniel-Mkpume, E.G. Okonkwo, V.S. Aigbodion, P.O. Offor, K.C. Nnakwo, Silica sand modified aluminium composite : An empirical study of the physical, mechanical and morphological properties, Mater. Res. Express. 6 (2019) 2–11. https://doi.org/doi.org/10.1088/2053-1591/ab14c6

[2] C.C. Daniel-Mkpume, C. Ugochukwu, E.G. Okonkwo, O.S.I. Fayomi, S.M. Obiorah, Effect of Luffa cylindrica fiber and particulate on the mechanical properties of epoxy, Int. J. Adv. Manuf. Technol. 101 (2019) 3439–3444. https://doi.org/https://doi.org/10.1007/s00170-019-03422-w

[3] S.I. Durowaye, G.I. Lawal, O.I. Sekunowo, E.G. Okonkwo, Synthesis and characterisation of hybrid polyethylene terephthalate matrix composites reinforced with Entada Mannii fibre particles and almond shell particles, J. King Saud Univ. - Eng. Sci. (2017) 1–9. https://doi.org/10.1016/j.jksues.2017.09.006

[4] B.C. Kandpal, R. Chaurasia, V. Khurana, Recent Advances in Green Composites – A Review, Int. J. Technol. Res. Eng. 2 (2015) 742–747

[5] N.L. García, L. Ribba, A. Dufresne, M.I. Aranguren, S. Goyanes, Physico-Mechanical Properties of Biodegradable Starch Nanocomposites, Macromol. Mater. Eng. 294 (2009) 169–177. https://doi.org/10.1002/mame.200800271

[6] A. Dufresne, J. Castano, Polysaccharide nanomaterial reinforced starch nanocomposites : A review, Starch. 68 (2016) 1–19. https://doi.org/10.1002/star.201500307

[7] R. Shanks, I. Kong, Thermoplastic Starch, in: Thermoplast. Elastomers, 2012: pp. 95–116

[8] M.L. Tanaka, N. Vest, C.M. Ferguson, P. Gatenholm, Comparison of Biomechanical Properties of Native Menisci and Bacterial Cellulose Implant, Int. J. Polym. Mater. Polym. Biomater. 63 (2014) 891–897. https://doi.org/10.1080/00914037.2014.886226

[9] J.M. Rajwade, K.M. Paknikar, J. V Kumbhar, Applications of bacterial cellulose and its composites in biomedicine, Appl Microbiol Biotechnol. (2015) 1–21. https://doi.org/10.1007/s00253-015-6426-3

[10] A. Svensson, E. Nicklasson, T. Harrah, B. Panilaitis, D.L. Kaplan, M. Brittberg, P. Gatenholm, Bacterial cellulose as a potential scaffold for tissue engineering of cartilage, Biomaterials. 26 (2005) 419–431. https://doi.org/10.1016/j.biomaterials.2004.02.049

[11] N.H. Zakaria, N. Muhammad, M.M.A.B. Abdullah, Potential of Starch Nanocomposites for Biomedical Applications, IOP Conf. Ser. Mater. Sci. Eng. 209 (2017) 1–7. https://doi.org/10.1088/1757-899X/209/1/012087

[12] O.A. El Seoud, H. Nawaz, E.P.G. Arêas, Chemistry and Applications of Polysaccharide Solutions in Strong Electrolytes/Dipolar Aprotic Solvents: An Overview, Molecules. 18 (2013) 1270–1313. https://doi.org/10.3390/molecules18011270

[13] R. Colussi, V. Zanella, S. Lisie, M. El, B. Biduski, L. Prietto, D. Dufech, R. Zavareze, A. Renato, G. Dias, Acetylated rice starches films with different levels of amylose : Mechanical , water vapor barrier , thermal , and biodegradability properties, Food Chem. 221 (2017) 1614–1620. https://doi.org/10.1016/j.foodchem.2016.10.129

[14] S.T. Sam, M.A. Nuradibah, K.M. Chin, N. Hani, Current Application and Challenges on Packaging Industry Based on Natural Polymer Blending, in: Nat. Polym., 2016: pp. 163–184. https://doi.org/10.1007/978-3-319-26414-1

[15] R.F.N. Quadrado, A.R. Fajardo, Microparticles based on carboxymethyl starch / chitosan polyelectrolyte complex as vehicles for drug delivery systems, Arab. J. Chem. (2018) 1–12. https://doi.org/10.1016/j.arabjc.2018.04.004

[16] H. Cornell, The Functionality of wheat starch, in: Starch Food, n.d.: pp. 211–240

[17] S. Yoon, Y. Deng, Clay – Starch Composites and Their Application in Papermaking, J. Appl. Polym. Sci. 100 (2006) 1032–1038. https://doi.org/10.1002/app.23007

[18] V.S. Waghmare, R.P. Wadke, S. Dyawanapelly, A. Deshpande, J. Ratnesh, P. Dandekar, Starch based nanofibrous scaffolds for wound healing applications, Bioact. Mater. 3 (2018) 255–266. https://doi.org/10.1016/j.bioactmat.2017.11.006

[19] A. Al-muhtaseb, W.A.. Mcminn, T.R.. Magee, Shrinkage , density and porosity variations during the convective drying of potato starch gel, in: Dry. 2004, 2004: pp. 1604–1611

[20] V.S. Naik, S.S. Prabhakara, S. Raghavendra, Mechanical properties of sisal fiber reinforced starch based bio composites, AIP Conf. Proc. 020019 (2019) 1–7.

https://doi.org/10.1063/1.5085590

[21] N. Singh, J. Singh, L. Kaur, N.S. Sodhi, B.S. Gill, Morphological , thermal and rheological properties of starches from different botanical sources, Food Chem. 81 (2003) 219–231

[22] M. Sujka, U. Pankiewicz, R. Kowalski, K. Nowosad, A. Noszczyk-nowak, Porous starch and its application in drug delivery systems, Polim Med. 48 (2018) 25–29. https://doi.org/10.17219/pim/99799

[23] N.H. Zakaria, N. Muhammad, M.M.A.B. Abdullah, I.G. Sandu, C.L. Mei Wan, Characteristics of Thermoplastic Potato Starch / Bentonite Nanocomposite Film, IOP Conf. Ser. Mater. Sci. Eng. 374 (2018) 1–5. https://doi.org/10.1088/1757-899X/374/1/012025

[24] A. Linton, K. Cato, T. Huang, Y. Chang, J. Ciou, J. Chang, H. Lin, Functional properties of arrowroot starch in cassava and sweet potato composite starches, Food Hydrocoll. 53 (2016) 187–191. https://doi.org/10.1016/j.foodhyd.2015.01.024

[25] A.H.. Abdullah, S. Chalimah, I. Primadona, M.H.. Hanantyo, Physical and chemical properties of corn , cassava , and potato starchs, IOP Conf. Ser. Earth Environ. Sci. 160 (2018) 1–6

[26] B. Jiang, S. Li, Y. Wu, J. Song, S. Chen, X. Li, H. Sun, Preparation and characterization of natural corn starch-based composite films reinforced by eggshell powder, CyTA - J. Food. 16 (2018) 1045–1054. https://doi.org/10.1080/19476337.2018.1527783

[27] A. Amri, I. Ekawati, S. Herman, S.. Yenti, Zultiniar, Y. Aziz, S.. Utami, Bahruddin, Properties enhancement of cassava starch based bioplastics with addition of graphene oxide, IOP Conf. Ser. Mater. Sci. Eng. 345 (2018) 1–13. https://doi.org/10.1088/1757-899X/345/1/012025

[28] K. Vaezi, G. Asadpour, H. Sharifi, Effect of ZnO nanoparticles on the mechanical, barrier and optical properties of thermoplastic cationic starch/montmorillonite biodegradable films, Int. J. Biol. Macromol. (2018) 1–36. https://doi.org/10.1016/j.ijbiomac.2018.11.142

[29] S. Naderizadeh, A. Shakeri, H. Mahdavi, N. Nikfarjam, N.T. Qazvini, Hybrid Nanocomposite Films of Starch , Poly (vinyl alcohol) (PVA), Starch Nanocrystals (SNCs), and Montmorillonite (Na-MMT): Structure – Properties Relationship, Starch. 7 (2019) 1–8. https://doi.org/10.1002/star.201800027

[30] C.L. Luchese, J.C. Spada, I.C. Tessaro, Starch content affects physicochemical properties of corn and cassava starch-based films, Ind. Crop. Prod. 109 (2017) 619–626. https://doi.org/10.1016/j.indcrop.2017.09.020

[31] P.R. Chang, P. Zheng, B. Liu, D.P. Anderson, J. Yu, X. Ma, Characterization of magnetic soluble starch-functionalized carbon nanotubes and its application for the adsorption of the dyes, J. Hazard. Mater. 186 (2011) 2144–2150. https://doi.org/10.1016/j.jhazmat.2010.12.119

[32] F.C. Soares, F. Yamashita, C.M.O. Müller, A.T.N. Pires, Effect of cooling and coating on thermoplastic starch / poly (lactic acid) blend sheets, Polym. Test. 33 (2014) 34–39. https://doi.org/10.1016/j.polymertesting.2013.11.001

[33] S. Mishra, T. Rai, Morphology and functional properties of corn , potato and tapioca starches, Food Hydrocoll. 20 (2006) 557–566. https://doi.org/10.1016/j.foodhyd.2005.01.001

[34] E.D.M. Teixeira, A.A.S. Curvelo, A.C. Corrêa, J.M. Marconcini, G.M. Glenn, L.H.C. Mattoso, Properties of thermoplastic starch from cassava bagasse and cassava starch and their blends with poly (lactic acid), Ind. Crop. Prod. 37 (2012) 61–68. https://doi.org/10.1016/j.indcrop.2011.11.036

[35] M. Kaur, S. Singh, Physicochemical , Morphological , Pasting , and Rheological Properties of Tamarind (Tamarindus indica L .) Kernel Starch, Int. J. Food Prop. 19 (2016) 2432–2442. https://doi.org/10.1080/10942912.2015.1121495

[36] N.Z. Ngobese, O.C. Wokadala, B. Du Plessis, L.S. Da Silva, A. Hall, S.P. Lepule, M. Penter, M.E.K. Ngcobo, H.C. Swart, Physicochemical and morphological properties of a small granule legume starch with atypical properties from wild mango (Cordyla africana L.) seeds: A comparison to maize, pea and kidney bean starch, (2018) 1–29. https://doi.org/10.1002/star.201700345

[37] R. Hoover, Composition , molecular structure , and physicochemical properties of tuber and root starches : a review, Carbohydr. Polym. 45 (2001) 253–267

[38] L. Li, H. Chen, M. Wang, X. Lv, Y. Zhao, L. Xia, Development and characterization of irradiated-corn-starch film, Carbohydr. Polym. (2018) 1–24. https://doi.org/10.1016/j.carbpol.2018.04.060

[39] L. Ren, X. Yan, J. Zhou, J. Tong, X. Su, Influence of chitosan concentration on mechanical and barrier properties of corn starch / chitosan films, Int. J. Biol. Macromol. 105 (2017) 1636–1643. https://doi.org/10.1016/j.ijbiomac.2017.02.008

[40] V. Vamadevan, E. Bertoft, Structure-function relationships of starch components,

Starch. 67 (2014) 55–68. https://doi.org/10.1002/star.201400188

[41] V.S. Aigbodion, E.G. Okonkwo, E.T. Akinlabi, Eco-friendly Polymer Composite : State-of-Arts , Opportunities and Challenge, in: Sustain. Polym. Compos. Nanocomposite, 2019: pp. 1233–1265

[42] M.K. Dang, R. Yoksan, Morphological characteristics and barrier properties of thermoplastic starch / chitosan blown film, Carbohydr. Polym. 150 (2016) 40–47. https://doi.org/10.1016/j.carbpol.2016.04.113

[43] Y. Zhong, X. Song, Y. Li, Antimicrobial , physical and mechanical properties of kudzu starch – chitosan composite films as a function of acid solvent types, Carbohydr. Polym. 84 (2011) 335–342. https://doi.org/10.1016/j.carbpol.2010.11.041

[44] P. Gonzalez-Seligra, L. Guz, O. Ochoa-Yepes, S. Goyanes, L. Fama, Influence of extrusion process conditions on starch film morphology, LWT - Food Sci. Technol. 84 (2017) 520–528. https://doi.org/10.1016/j.lwt.2017.06.027

[45] M. Combrzyński, L. Mościcki, M. Mitrus, K. Kupryaniuk, A. Oniszczuk, Application of extrusion-cooking technique for foamed starch-based materials, Contemp. Res. Trends Agric. Eng. 10 (2018) 1–6

[46] L. Moscicki, M. Mitrus, A. Wojtowicz, T. Oniszczuk, A. Rejak, Extrusion-Cooking of Starch, in: Adv. Agrophysical Res., 2013: pp. 319–346

[47] C. López de Dicastillo, K. Roa, L. Garrido, A. Pereira, M.J. Galotto, Novel Polyvinyl Alcohol/Starch Electrospun Fibers as a Strategy to Disperse Cellulose Nanocrystals into Poly(lactic acid), Polymers (Basel). 9 (2017) 1–16. https://doi.org/10.3390/polym9040117

[48] H. Kim, S.S. Park, S. Lim, Preparation, characterization and utilization of starch nanoparticles, Colloids Surfaces B Biointerfaces. (2014) 1–14. https://doi.org/10.1016/j.colsurfb.2014.11.011

[49] N. Soykeabkaew, C. Thanomsilp, O. Suwantong, A review: starch-based composite foams, Compos. PART A. 78 (2015) 246–263. https://doi.org/10.1016/j.compositesa.2015.08.014

[50] J.F. Mendes, R.T. Paschoalin, V.B. Carmona, A.R. Sena Neto, A.C.P. Marques, J.M. Marconcini, L.H.C. Mattoso, E.S. Medeiros, J.E. Oliveira, Biodegradable polymer blends based on corn starch and thermoplastic chitosan processed by extrusion, Carbohydr. Polym. 137 (2016) 452–458

[51] L.A. Toro-márquez, D. Merino, T.J. Gutiérrez, Bionanocomposite Films Prepared

from Corn Starch With and Without Nanopackaged Jamaica (Hibiscus sabdariffa) Flower Extract, Food Bioprocess Technol. (2018) 1–19. https://doi.org/10.1007/s11947-018-2160-z

[52] R.P.H. Brandelero, M.V.E. Grossmann, F. Yamashita, Effect of the method of production of the blends on mechanical and structural properties of biodegradable starch films produced by blown extrusion, Carbohydr. Polym. 86 (2011) 1344–1350. https://doi.org/10.1016/j.carbpol.2011.06.045

[53] L. Quiles-Carrillo, N. Montanes, F. Pineiro, A. Jorda-Vilaplana, S. Torres-Giner, Ductility and Toughness Improvement of Injection-Molded Compostable Pieces of Polylactide by Melt Blending with Poly(ε-caprolactone) and Thermoplastic Starch, Materials (Basel). 11 (2018) 1–20. https://doi.org/10.3390/ma11112138

[54] C.M.O. Müller, J. Borges, F. Yamashita, Composites of thermoplastic starch and nanoclays produced by extrusion and thermopressing, Carbohydr. Polym. 89 (2012) 504–510. https://doi.org/10.1016/j.carbpol.2012.03.035

[55] J. Bonilla, L. Atarés, M. Vargas, A. Chiralt, Properties of wheat starch film-forming dispersions and films as affected by chitosan addition, J. Food Eng. 114 (2013) 303–312. https://doi.org/10.1016/j.jfoodeng.2012.08.005

[56] N.H. Zakaria, N. Muhammad, A. V Sandu, M.M.A.B. Abdullah, Effect of Mixing Temperature on Characteristics of Thermoplastic Potato Starch Film, IOP Conf. Ser. Mater. Sci. Eng. 374 (2018) 1–6. https://doi.org/10.1088/1757-899X/374/1/012083

[57] A.L. Da Róz, M.D. Zambon, A.A.S. Curvelo, A.J.F. Carvalho, Thermoplastic starch modified during melt processing with organic acids : The effect of molar mass on thermal and mechanical properties, Ind. Crops Prod. 33 (2011) 152–157. https://doi.org/10.1016/j.indcrop.2010.09.015

[58] M. Thunwall, A. Boldizar, M. Rigdahl, Compression Molding and Tensile Properties of Thermoplastic Potato Starch Materials, Biomacromolecules. 7 (2006) 981–986

[59] R. Ortega-toro, I. Morey, P. Talens, A. Chiralt, Active bilayer films of thermoplastic starch and polycaprolactone obtained by compression molding, Carbohydr. Polym. 127 (2015) 282–290. https://doi.org/10.1016/j.carbpol.2015.03.080

[60] O. V Lopez, F. Versino, M.A. Villar, M.A. Garcia, Agro-industrial residue from starch extraction of Pachyrhizus ahipa as filler of thermoplastic corn starch films,

Materials Research Forum LLC
https://doi.org/10.21741/9781644900772-7

Carbohydr. Polym. 134 (2015) 324–332.
https://doi.org/10.1016/j.carbpol.2015.07.081

[61] Z. Liu, J. He, Polyvinyl alcohol / starch composite nanofibers by bubble electrospinning, Therm. Sci. 18 (2014) 1473–1475. https://doi.org/10.2298/TSCI1405473L

[62] Y. Wu, M. Ji, Q. Qi, Y. Wang, L. Zhang, Preparation , Structure , and Properties of Starch / Rubber Composites Prepared by Co-Coagulating Rubber Latex and Starch Paste, Macromol. Rapid Commun. 25 (2004) 565–570. https://doi.org/10.1002/marc.200300125

[63] R.J. Wanter, D. Mangindaan, Preparation of Indonesian taro starch particles via precipitation process, IOP Conf. Ser. Earth Environ. Sci. 195 (2018) 1–3. https://doi.org/10.1088/1755-1315/195/1/012059

[64] A.M. Amini, S. Mohammad, A. Razavi, S.A. Mortazavi, Morphological , physicochemical , and viscoelastic properties of sonicated corn starch, Carbohydr. Polym. 122 (2015) 282–292. https://doi.org/10.1016/j.carbpol.2015.01.020

[65] F. Askari, E. Sadeghi, R. Mohammadi, M. Rouhi, M. Taghizadeh, M.H. Shirgardoun, M. Kariminejad, The physicochemical and structural properties of psyllium gum / modified starch composite edible film, J. Food Process. Preserv. 42 (2018) 1–9. https://doi.org/10.1111/jfpp.13715

[66] H. Li, X. Gao, Y. Wang, X. Zhang, Z. Tong, Comparison of chitosan / starch composite film properties before and after ceoss-linking, Int. J. Biol. Macromol. 52 (2013) 275–279. https://doi.org/10.1016/j.ijbiomac.2012.10.016

[67] H.P.S. Abdul Khalil, S.W. Yap, F.A.T. Owolabi, M.K.M. Haafiz, M.R. Fazita, D.A. Gopakumar, M. Hasan, S. Rizal, Techno-functional Properties of Edible Packaging Films at Different Polysaccharide Blends, J. Phys. Sci. 30 (2019) 23–41

[68] R.A. Garalde, R. Thipmanee, P. Jariyasakoolroj, A. Sane, The effects of blend ratio and storage time on thermoplastic starch / poly (butylene adipate- co -terephthalate) films, Heliyon. 5 (2019) 1–20. https://doi.org/10.1016/j.heliyon.2019.e01251

[69] K. Katerinopoulou, A. Giannakas, N. Barkoula, A. Ladavos, Preparation , characterization and biodegradability assessment of Maize starch-(PVOH)/Clay nanocomposite films, Starch. 71 (2019) 1–8. https://doi.org/10.1002/star.201800076

[70] A. More, C. Sen, M. Das, A Comparative Study of Glutaraldehyde and Citric Acid As a Crosslinking Agent in Starch- Polyvinyl Alcohol Based Biodegradable Film,

in: Int. Conf. Emerg. Technol. Agric. Eng. IIT Kharagpur, 2019: pp. 1–6

[71] R. Kumar, G. Ghoshal, M. Goyal, Synthesis and functional properties of gelatin / CA – starch composite film : excellent food packaging material Synthesis and functional properties of gelatin / CA – starch composite film : excellent food packaging material, J. Food Sci. Technol. 56 (2019) 1954–1965. https://doi.org/10.1007/s13197-019-03662-4

[72] S. Kumar, S.K. Samal, S. Mohanty, S.K. Nayak, Synthesis and Characterization of Nanoclay-Reinforced Trifunctional " Bioresin-Modified " Epoxy Blends Enhanced with Mechanical and Thermal Properties, ChemistrySelect. 2 (2017) 11445–11455. https://doi.org/10.1002/slct.201702041

[73] P.S. Garcia, F.R.B. Turbiani, A.M. Baron, G.L. Brizola, M.A. Tavares, F. Yamashita, D. Eiras, M.V.E. Grossmann, Sericin as compatibilizer in starch / polyester blown films, Polimeros. (2018) 1–6. https://doi.org/https://doi.org/10.1590/0104-1428.05117

[74] S. Sukhija, S. Singh, C.S. Riar, Development and characterization of biodegradable films from whey protein concentrate, psyllium husk and oxidized, crosslinked, dual-modified lotus rhizome starch composite, J. Sci. Food Agric. 99 (2019) 3398–3409. https://doi.org/10.1002/jsfa.9557

[75] M.G.A. Vieira, M. Altenhofen da Silva, L. Oliveira dos Santos, M.M. Beppu, Natural-based plasticizers and biopolymer films : A review, Eur. Polym. J. 47 (2011) 254–263. https://doi.org/10.1016/j.eurpolymj.2010.12.011

[76] G. Madhumitha, J. Fowsiya, S.M. Roopan, V.K. Thakur, Recent advances in starch – clay nanocomposites, Int. J. Polym. Anal. Charact. 23 (2018) 331–345. https://doi.org/10.1080/1023666X.2018.1447260

[77] S. Ismail, N. Mansor, Z. Man, A Study on Thermal Behaviour of Thermoplastic Starch Plasticized by [Emim] Ac and by [Emim] Cl, Procedia Eng. 184 (2017) 567–572. https://doi.org/10.1016/j.proeng.2017.04.138

[78] N.H. Zakaria, N. Muhammad, M. Mustafa, A. Bakri, Effect of glycerol content on mechanical , microstructure and physical properties of thermoplastic potato starch film, AIP Conf. Proc. 2030 (2018) 1–5. https://doi.org/10.1063/1.5066871

[79] A. Ekielski, T. Zelanzinski, V. Vladut, E. Tulska, Impact of bioethanol additive on the properties of stretchable starch films, Ann. Warsaw Univ. Life Sci. – SGGW. 69 (2017) 79–88. https://doi.org/10.22630/AAFE.2017.9

[80] A.H. Mohd Zain, A.W.M. Kahar, N.Z. Noriman, Chemical-Mechanical Hydrolysis

Technique of Modified Thermoplastic Starch for Better Mechanical Performance, Procedia Chem. 19 (2016) 638–645. https://doi.org/10.1016/j.proche.2016.03.064

[81] Y. Wang, M.A. Rodriguez-Perez, R.L. Reis, J.F. Mano, Thermal and Thermomechanical Behaviour of Polycaprolactone and Starch / Polycaprolactone Blends for Biomedical Applications, Macromol. Mater. Eng. 290 (2005) 792–801. https://doi.org/10.1002/mame.200500003

[82] M.L. Sanyang, R.A. Ilyas, S.M. Sapuan, R. Jumaidin, Sugar Palm Starch-Based Composites for Packaging Applications, in: Bionanocomposites Packag. Appl., 2018: pp. 125–147. https://doi.org/10.1007/978-3-319-67319-6

[83] D. Lin, Y. Kuang, G. Chen, Q. Kuang, C. Wang, P. Zhu, C. Peng, Z. Fang, Enhancing moisture resistance of starch-coated paper by improving the film forming capability of starch film, Ind. Crop. Prod. 100 (2017) 12–18. https://doi.org/10.1016/j.indcrop.2017.02.013

[84] C. Liza, B. Soegijono, E. Budianto, Syuhada, D. Rusmana, Photodegradation Effect of Structure of Linear Low Density Polyethylene – Starch – Clay Nanocomposite Film, IOP Conf. Ser. Mater. Sci. Eng. 395 (2018) 1–7. https://doi.org/10.1088/1757-899X/395/1/012019

[85] O. Moreno, L. Atarés, A. Chiralt, Effect of the incorporation of antimicrobial / antioxidant proteins on the properties of potato starch films, Carbohydr. Polym. 133 (2015) 353–364. https://doi.org/10.1016/j.carbpol.2015.07.047

[86] S.C. Mendes, R.L. Reis, Y.P. Bovell, A.M. Cunha, C.A. Van Blitterswijk, J.D. De Bruijn, Biocompatibility testing of novel starch-based materials with potential application in orthopaedic surgery : a preliminary study, Biomaterials. 22 (2001) 2057–2064

[87] S. Lefnaoui, N. Moulai-mostefa, Synthesis and evaluation of the structural and physicochemical properties of carboxymethyl pregelatinized starch as a pharmaceutical excipient, Saudi Pharm. J. 23 (2015) 698–711. https://doi.org/10.1016/j.jsps.2015.01.021

[88] A. Rodrigues, M. Emeje, Recent applications of starch derivatives in nanodrug delivery, Carbohydr. Polym. 87 (2012) 987–994. https://doi.org/10.1016/j.carbpol.2011.09.044

[89] S.H.M. Najafi, M. Baghaie, A. Ashori, Preparation and characterization of acetylated starch nanoparticles as drug carrier : Ciprofloxacin as a model, Int. J. Biol. Macromol. 87 (2016) 48–54. https://doi.org/10.1016/j.ijbiomac.2016.02.030

[90] C. Calinescu, B. Mondovi, R. Federico, P. Ispas-szabo, M.A. Mateescu, Carboxymethyl starch : Chitosan monolithic matrices containing diamine oxidase and catalase for intestinal delivery, Int. J. Pharm. 428 (2012) 48–56. https://doi.org/10.1016/j.ijpharm.2012.02.032

[91] H. Xiao, T. Yang, Q. Lin, G. Liu, L. Zhang, F. Yu, Y. Chen, Acetylated starch nanocrystals : Preparation and antitumor drug delivery study, Int. J. Biol. Macromol. 89 (2016) 456–464. https://doi.org/10.1016/j.ijbiomac.2016.04.037

[92] Y. Li, X. Zhao, L. Wang, Y. Liu, W. Wu, C. Zhong, Q. Zhang, J. Yang, Preparation , characterization and in vitro evaluation of melatonin-loaded porous starch for enhanced bioavailability, Carbohydr. Polym. 202 (2018) 125–133. https://doi.org/10.1016/j.carbpol.2018.08.127

[93] Y. Benavent-gil, C.M. Rosell, Morphological and physicochemical characterization of porous starches obtained from different botanical sources and amylolytic enzymes, Int. J. Biol. Macromol. 103 (2017) 587–595. https://doi.org/10.1016/j.ijbiomac.2017.05.089

[94] C. Wu, Z. Wang, Z. Zhi, T. Jiang, J. Zhang, S. Wang, Development of biodegradable porous starch foam for improving oral delivery of poorly water soluble drugs, Int. J. Pharm. 403 (2011) 162–169. https://doi.org/10.1016/j.ijpharm.2010.09.040

[95] M.J. Santander-ortega, T. Stauner, B. Loretz, J.L. Ortega-vinuesa, D. Bastos-gonzález, G. Wenz, U.F. Schaefer, C.M. Lehr, Nanoparticles made from novel starch derivatives for transdermal drug delivery, J. Control. Release. 141 (2010) 85–92. https://doi.org/10.1016/j.jconrel.2009.08.012

[96] C. Belingheri, A. Ferrillo, E. Vittadini, Porous starch for flavor delivery in a tomato-based food application, LWT - Food Sci. Technol. 60 (2015) 593–597. https://doi.org/10.1016/j.lwt.2014.09.047

[97] G. Gong, F. Zhang, Z. Cheng, L. Zhou, Facile fabrication of magnetic carboxymethyl starch / poly (vinyl alcohol) composite gel for methylene blue removal, Int. J. Biol. Macromol. 81 (2015) 205–211. https://doi.org/10.1016/j.ijbiomac.2015.07.061

[98] L. Yan, P.R. Chang, P. Zheng, Preparation and characterization of starch-grafted multiwall carbon nanotube composites, Carbohydr. Polym. 84 (2011) 1378–1383. https://doi.org/10.1016/j.carbpol.2011.01.042

[99] M. Zubair, N. Jarrah, Ihsanullah, A. Khalid, S.M. Manzar, T.S. Kazeem, M.A. Al-

Harthi, Starch-NiFe-layered double hydroxide composites : Efficient removal of methyl orange from aqueous phase, J. Mol. Liq. 249 (2018) 254–264. https://doi.org/10.1016/j.molliq.2017.11.022

[100] C. Yu, X. Tang, S. Liu, Y. Yang, X. Shen, C. Gao, Laponite crosslinked starch / polyvinyl alcohol hydrogels by freezing / thawing process and studying their cadmium ion absorption, Int. J. Biol. Macromol. 117 (2018) 1–6. https://doi.org/10.1016/j.ijbiomac.2018.05.159

[101] J. Wang, S. Yuan, Y. Wang, H. Yu, Synthesis , characterization and application of a novel starch- based flocculant with high flocculation and dewatering properties, Water Res. (2013) 1–6. https://doi.org/10.1016/j.watres.2013.01.050

Keyword Index

About the Editors

Dr. Amir Al-Ahmed is working as a Research Scientist-II (Associate Professor) in the Center of Research Excellence in Renewable Energy, at King Fahd University of Petroleum & Minerals (KFUPM), Saudi Arabia. He graduated in chemistry from Aligarh Muslim University (AMU), India. He obtained his M.Phil (2001) and Ph.D. (2003) degree in Applied Chemistry on conducting polymer based composites and its applications, from the Zakir Hussain College of Engineering and Technology, AMU, India. During his postdoctoral research activity, he worked on different multi-disciplinary project in South Africa and Saudi Arabia such as biological and chemical sensor, energy storage and conversion and also on CO_2 reduction. Throughout his academic career, he has gained extensive experience in materials chemistry and electrocatalysis for applications in energy conversion, storage, sensors and membranes. Currently, he is involved in several multidisciplinary research projects, funded by Saudi national research programs (thin film solar cells) and Saudi Aramco. Dr. Amir has on his credit a good number of articles published in international scientific journals, conferences and as book chapters. He has edited six books (publisher: Trans Tech Publication, Switzerland) and he is in the process of editing and writing another two books with Springer and Elsevier. He is also the Editor-in-Chief of an international journal "Nano Hybrids and Composites" along with Professor Y.H. Kim.

Dr. Inamuddin is currently working as Assistant Professor in the Chemistry Department, Faculty of Science, King Abdulaziz University, Jeddah, Saudi Arabia. He is a permanent faculty member (Assistant Professor) at the Department of Applied Chemistry, Aligarh Muslim University, Aligarh, India. He obtained Master of Science degree in Organic Chemistry from Chaudhary Charan Singh (CCS) University, Meerut, India, in 2002. He received his Master of Philosophy and Doctor of Philosophy degrees in Applied Chemistry from Aligarh Muslim University (AMU), India, in 2004 and 2007, respectively. He has extensive research experience in multidisciplinary fields of Analytical Chemistry, Materials Chemistry, and Electrochemistry and, more specifically, Renewable Energy and Environment. He has worked on different research projects as project fellow and senior research fellow funded by University Grants Commission (UGC), Government of India, and Council of Scientific and Industrial Research (CSIR), Government of India. He has received Fast Track Young Scientist Award from the Department of Science and Technology, India, to work in the area of bending actuators and artificial muscles. He has completed four major research projects sanctioned by University Grant Commission, Department of Science and Technology, Council of

Scientific and Industrial Research, and Council of Science and Technology, India. He has published 162 research articles in international journals of repute and eighteen book chapters in knowledge-based book editions published by renowned international publishers. He has published 80 edited books with Springer (U.K.), Elsevier, Nova Science Publishers, Inc. (U.S.A.), CRC Press Taylor & Francis Asia Pacific, Trans Tech Publications Ltd. (Switzerland), IntechOpen Limited (U.K.), and Materials Research Forum LLC (U.S.A). He is a member of various journals' editorial boards. He is also serving as Associate Editor for journals (Environmental Chemistry Letter, Applied Water Science and Euro-Mediterranean Journal for Environmental Integration, Springer-Nature), Frontiers Section Editor (Current Analytical Chemistry, Bentham Science Publishers), Editorial Board Member (Scientific Reports-Nature), Editor (Eurasian Journal of Analytical Chemistry), and Review Editor (Frontiers in Chemistry, Frontiers, U.K.). He is also guest-editing various special thematic special issues to the journals of Elsevier, Bentham Science Publishers, and John Wiley & Sons, Inc. He has attended as well as chaired sessions in various international and national conferences. He has worked as a Postdoctoral Fellow, leading a research team at the Creative Research Initiative Center for Bio-Artificial Muscle, Hanyang University, South Korea, in the field of renewable energy, especially biofuel cells. He has also worked as a Postdoctoral Fellow at the Center of Research Excellence in Renewable Energy, King Fahd University of Petroleum and Minerals, Saudi Arabia, in the field of polymer electrolyte membrane fuel cells and computational fluid dynamics of polymer electrolyte membrane fuel cells. He is a life member of the Journal of the Indian Chemical Society. His research interest includes ion exchange materials, a sensor for heavy metal ions, biofuel cells, supercapacitors and bending actuators.

www.ingramcontent.com/pod-product-compliance
Lightning Source LLC
Chambersburg PA
CBHW071157210326
41597CB00016B/1584